T0345140

Technology, Human Performance, and Nuclear Facilities

This book statistically confirms that complexity and changing technologies that affect the way operators interact within the systems of the nuclear facilities exacerbate the severity of incidents caused by human error and details the application of the systems engineering process to reduce human error given industries' rapidly advancing technology.

Technology, Human Performance, and Nuclear Facilities: A Systems Engineering Approach to Reduce Human Error provides a basic understanding of Human Error/Performance and its relation to industrial operations and advancing technologies incorporated into facilities. The book discusses the context surrounding the complexity of changing technologies at nuclear facilities and the potential worsening of problems caused by human error when technology advancements concerning operator interaction with control systems are implemented. It presents how to reduce human error propensity given the incorporation of advanced technology and covers ways to reduce human error using the systems engineering process. Also offered are several concepts related to the operator's involvement in the systems engineering process and the human performance integration with system operational requirements and system testing, evaluation, and validation, and the procedures and training development in the systems engineering process. This book presents empirical evidence for the importance of human performance management in the context of nuclear facilities and offers practical recommendations for the improvement of this function.

Systems engineers, plant/design engineers, the nuclear industry, plant operations management, and those involved in industrial and nuclear safety will find something of interest in this book.

Technology, Human Performance, and Nuclear Facilities

A Systems Engineering Approach to Reduce Human Error

Jonathan K. Corrado

CRC Press is an imprint of the
Taylor & Francis Group, an **informa** business

First edition published 2023
by CRC Press
6000 Broken Sound Parkway NW, Suite 300, Boca Raton, FL 33487-2742

and by CRC Press
4 Park Square, Milton Park, Abingdon, Oxon, OX14 4RN

CRC Press is an imprint of Taylor & Francis Group, LLC

Library of Congress Cataloging-in-Publication Data

Names: Corrado, Jonathan K., author.
Title: Technology, human performance, and nuclear facilities : a systems engineering approach to reduce human error / Jonathan K. Corrado.
Description: First edition. | Boca Raton : CRC Press, 2023. | Includes bibliographical references and index.
Identifiers: LCCN 2022031349 (print) | LCCN 2022031350 (ebook) | ISBN 9781032385501 (hbk) | ISBN 9781032386782 (pbk) | ISBN 9781003346265 (ebk)
Subjects: LCSH: Nuclear power plants—Human factors.
Classification: LCC TK9152.14 .C67 2023 (print) | LCC TK9152.14 (ebook) | DDC 629.134/355—dc23/eng/20220822
LC record available at https://lccn.loc.gov/2022031349
LC ebook record available at https://lccn.loc.gov/2022031350

ISBN: 978-1-032-38550-1 (hbk)
ISBN: 978-1-032-38678-2 (pbk)
ISBN: 978-1-003-34626-5 (ebk)

DOI: 10.1201/9781003346265

Typeset in Times
by codeMantra

This book is dedicated to Earth's systems engineer.

For in him all things were created: things in heaven and on earth, visible and invisible, whether thrones or powers or rulers or authorities; all things have been created through him and for him. He is before all things, and in him all things hold together – Colossians 1:16–17 (KJV)

Contents

Preface

Many unfortunate and unintended adverse industrial incidents occur across the United States each year, and the nuclear industry is no exception. Depending on the severity, these incidents can be problematic for people, the facilities, and the surrounding environments. These incidents occur for a number of varying reasons, but more often than not, human error is an accomplice.

In the course of this research, extensive benchmarking, evaluation, debate, and investigation revealed a key underlying hindrance to successful facility operation: a disproportionate dependence on "technology-only" enhancements. Technology is intended to make us more productive, but its use can carry a penalty. The computers, decision support systems, and complex control and logic programs used at plants can gradually diminish intuition and expertise and can ultimately become the replacement for a robust, knowledge-based training and Human Performance Improvement (HPI) program. Technology is a powerful tool, but in the operational setting, it must be properly balanced with thorough training and adherence to human error prevention techniques and conduct.

During my study of human error, I recognized that devising suitable steps to prevent human error is crucial in all aspects of a project and must permeate all phases of the systems engineering process. HPI is much more than employing a set of human performance tools. Human error psychology, effects, risks, error traps, and mental models must be examined, consciously applied, and woven into the operating structure of plant organizations, especially in light of the innate complexities associated with technological advancements. Therefore, this book seeks to present empirical evidence for the importance of human performance management in the context of nuclear facilities and to offer practical recommendations for the improvement of this function.

The content of this book stems from my experience in the nuclear Navy and nuclear industry in addition to drawing from my doctoral dissertation:

Corrado, Jonathan. 2017. "Technological Advances, Human Performance, and the Operation of Nuclear Facilities," PhD dissertation, Colorado State University, Fort Collins, CO, ProQuest (AAT 10258407).

Acknowledgments

Thank you to the American Society of Mechanical Engineers for republication permission (conveyed through Copyright Clearance Center, Inc.) for content from these articles:

Corrado, Jonathan, and Ronald Sega. 2020. "Impact of Advancing Technology on Nuclear Facility Operation," *ASCE-ASME Journal of Risk and Uncertainty in Engineering Systems, Part B: Mechanical Engineering* 6(1). doi:10.1115/1.4044784.

Corrado, Jonathan. 2022. "Proactive Human Error Reduction Using the Systems Engineering Process," *ASME Journal of Nuclear Engineering and Radiation Science* 8(2). doi:10.1115/1.4051362.

Corrado, Jonathan. 2021. "The Intersection of Advancing Technology and Human Performance," *ASME Journal of Nuclear Engineering and Radiation Science* 7(1). doi:10.1115/1.4047717.

Corrado, Jonathan. 2022. "The Incorporation of Human Performance Improvement into Systems Design," *ASME Journal of Nuclear Engineering and Radiation Science* 8(2). doi:10.1115/1.4051792.

Corrado, Jonathan. 2022. "Survey Use to Validate Engineering Methodology and Enhance System Safety," *ASCE-ASME Journal of Risk and Uncertainty in Engineering Systems, Part B: Mechanical Engineering* 8(2). doi:10.1115/1.4053305.

Also, thanks to the Korean Nuclear Society for republication permission for content from:

Corrado, Jonathan. 2021. "Human-Machine System Optimization in Nuclear Facility Systems," *Nuclear Engineering and Technology* 53(10). doi:10.1016/j.net.2021.04.022.

Author

Jonathan K. Corrado's professional background is primarily in the nuclear industry, where he has expertise in program management, nuclear safety, nuclear criticality safety, regulatory compliance, event investigation and reporting, regulatory commitment management, and oversight of industrial safety, waste management, environmental monitoring, and Nuclear Materials Control and Accountability.

Prior to his work in the nuclear industry, Dr. Corrado served in the US Navy, where he was a nuclear surface warfare officer. He continued his naval career in the US Navy Reserve where he is currently a senior officer and has held a breadth of assignments including command roles on several occasions. Dr. Corrado also briefly worked at a DoD Laboratory and in the defense industry managing systems engineering.

Dr. Corrado holds a Bachelor of Science degree in Mechanical Engineering from the Virginia Military Institute, a Master of Engineering Management from the Old Dominion University, a PhD in Systems Engineering from the Colorado State University, and is a graduate of the Navy Nuclear Power Program and Naval Postgraduate School. He is also a licensed Professional Engineer in Mechanical Engineering in the State of Ohio.

He has several fields of research interest, including nuclear engineering, systems engineering, human performance, and military affairs, and has authored several refereed journal articles in these fields.

1 Incident Case Studies in the US Nuclear Power Industry and Impetus for Human Error Reduction

As of 2013, the United States has had roughly 440 nuclear power facilities that have been in operation for roughly over 14,700 reactor-years. During this time, there have been twenty-three reactor core meltdowns of varying severity in both commercial and research operations, or one major nuclear accident for every 640 reactor years. According to the international design requirements for nuclear facilities, a reactor core meltdown should only occur about once every 20,000 reactor-years. This means that the incidence of US reactor core meltdowns is thirty-two times higher than what theory would predict [1].

INCIDENT CASE STUDIES IN THE US NUCLEAR POWER INDUSTRY

Most Americans are familiar with the incident at Three Mile Island Nuclear Plant near Middletown, Pennsylvania, on March 28, 1979, primarily because nuclear power was relatively new and intimidating in 1979, it was the most

DOI: 10.1201/9781003346265-1

expensive nuclear accident in US history, and it attained a rating of 5 (out of 7) on the International Nuclear and Radiological Event Scale (INES).

The problem began with a pilot-operated relief valve that became stuck open within the primary system, allowing a significant amount of coolant from the reactor to leak from the core albeit contained within the reactor containment structure. The initial problem was a mechanical failure, but it was exacerbated by the plant operators' failure to recognize that the coolant was leaking. Control room indicators had recently been added to a new user interface for which the operators did not have adequate training. In fact, one of the operators was unaware of the placement of an indicator light and overrode the automatically operated emergency cooling system, believing an excess of coolant within the reactor was causing the release of steam pressure [2].

The incident at Three Mile Island led quickly to an increase in US government regulation of the nuclear industry [3]. The partial meltdown that resulted from the incident released radioactive iodine and gases of unknown quantities into the environment. Several epidemiologic studies have been conducted regarding the rates of cancer and other illnesses related to radiation in the area near the facility. Though no statistically significant increase in problems has been noted, the cleanup of the area lasted for fourteen years [4]. The Three Mile Island accident is especially important here because it had such a high cost (in terms of money), and multiple in-depth investigations of the accident took place [5], along with studies from the perspectives of human factors and user interface engineering.

What is not as well known to the general public is that while Three Mile Island may hold the record for expense, it is hardly alone in terms of the damage to the facility and the potential danger to human life it posed. Moreover, of the twenty-three nuclear reactor meltdowns from 1955 to 2021, eighteen were caused by some type of human error, and several of these also involved unfamiliarity with new equipment [6].

Indeed, a simple review of US nuclear incidents from 1955 to 2021 that reached Level 3 (serious incident) or higher on the INES shows that the cost of each incident at facilities that had recently undergone technological changes affecting plant operators' jobs was higher than facilities that had not undergone changes. Moreover, the incorporation of new technologies into nuclear facilities and spending more money on upgrades increased both the facility's capacity and the number of incidents reported.

Breaking these incidents down by case is even more enlightening. For example, the reactor in Simi Valley, California, the first nuclear reactor to provide electricity in the United States, experienced a partial core meltdown on July 26, 1959, due to human error [7]. The entire facility was an experiment in using a sodium reactor to produce electricity through nuclear power. Therefore, throughout the facility's operation, new instruments and technologies were introduced. The meltdown involved thirteen of forty-three fuel elements and the release of radioactive gas [7]. One reason for the partial meltdown is that the individuals working at the facility were using new equipment and gauges they were unfamiliar with. The same is true for the January 3, 1961, SL-1 experimental reactor meltdown at Idaho Falls [5]. Three people in the reactor room died due to exposure to radiation. In fact, 500 R per hour were still being emitted from the bodies when the rescue workers arrived. A rod ejection caused one body to be lodged in the ceiling of the reactor room, held there by the control rod that was launched during the reactor super-criticality. The individuals in the control room were so irradiated that they had to be buried in coffins made of lead [2]. These operators too were working with new equipment, including the gauges that indicated whether the control rods were in their proper location and orientation [4].

Human error leading to reactor incidents and involving new equipment also occurred on October 5, 1966, in the Enrico Fermi Nuclear Power Plant in Monroe, Michigan [7]. The primary cause of the increase in temperature was determined to be blockage of a spigot needed for the liquid sodium coolant to enter the reactor, causing its temperature to rise slowly over several hours [2]. Operators did not become aware of the problem until alarms sounded regarding the elevated core temperature, by which time, a partial fuel meltdown had already occurred. The new equipment was new gauges for reading the core temperature, and operator unfamiliarity with the instruments likely led to the temperature change being overlooked [4]. The next such incident was on March 20, 1982, at the Nine Mile Point Nuclear Generating Station in Scriba, New York, when the system piping failed in the recirculation system of unit one. As a result, the unit was closed for two years. The technology used in the plant was relatively new, and the failure of the recirculation piping was due to human error when engineering the facility [5]. On March 25, 1982, at the Indian Point Energy Center Nuclear Power Plant in Buchanan, New York, unit three was shut down because the steam generator tubes were damaged when the facility operators failed to notice warnings that it was not safe to use the section of the reactor that relied on these tubes [6]. The interface that they were using had been installed only a few weeks before the incident occurred, and the reactor personnel were not yet trained in their operation [4]. On June 9, 1985, the Davis-Besse facility in Oak Harbor, Ohio experienced a near-meltdown stemming from a shutdown of the primary feedwater pumps supplying water for the reactor's steam generators [8]. An operator in the control room tried to start the emergency feedwater pumps, but the emergency pumps entered an over-speed condition because of operator error. The human interface had been upgraded only one month prior to the incident [9].

Not all lethal or damaging incidents involve meltdowns, near-meltdowns, or extensive shutdowns, of course. Tragedy hit Wolf Creek Generating Station in Burlington, Kansas on July 15, 1987, when a safety inspector made contact with a mislabeled wire, was inadvertently electrocuted, and died [8]. The electrical system in this part of the facility had recently been upgraded, and the worker doing the labeling had not been properly advised of the new wiring system [6]. Another death occurred at the facility on March 29, 1988, when a worker fell into an unmarked manhole and was electrocuted while trying to escape [10]. The manhole had recently been added, and the workers failed to label the manhole properly. In addition, they were not familiar enough with the new system to understand the necessary safety measures to ensure that someone entering the manhole would not be electrocuted [9].

Beginning on September 10, 1988, unit two of the facility in Surry, Virginia, was shut down for a year [3] because of the failure of a seal on a refueling cavity, which resulted in the destruction of an internal pipe system. An investigation revealed that both the internal pipe system and the cavity seal had been recently upgraded, but that the maintenance and inspection personnel at the facility were not familiar with the new systems [9]. The Palo Verde Nuclear Generating Station near Tonopah, Arizona, saw a failure of the atmospheric dump valves, an associated transformer fire, and an emergency shutdown on March 5, 1989 [10]. Many of the maintenance and inspection personnel were not yet familiar with the facility's unique cooling system [3]. After a fire on November 17, 1991, the FitzPatrick Nuclear Reactor was shut down for over a year after a discovery that several safety and fire procedures were not being followed [3]. An investigation revealed that the fire and safety systems had recently been upgraded, but that the associated staff at the facility had not been properly trained on the new equipment [5]. The Brunswick reactor in Southport, North Carolina, shut down two units on April 21, 1992, after a failure in the emergency diesel generators who monitoring equipment had recently been upgraded. The facility personnel working with this new equipment stated that they did not understand how to properly interpret some of the readings. The auxiliary feedwater pumps at the South Texas Project's units one and two in Bay City, Texas, failed on February 3, 1993, causing both reactors to be shut down. The maintenance personnel were not properly trained on the newly upgraded reactor cooling systems. Multiple equipment failures and broken pipes led to the shutdown of unit one at the Sequoyah Nuclear Plant, located near Soddy-Daisy, Tennessee, on March 2, 1993. The cause was improper maintenance on newly installed equipment and operator error when reading the newly installed instruments [10].

Shutdowns caused by human error when dealing with new equipment have occurred at least four more times since then, including incidents at the Dresden Generating Station near Morris, Illinois, on May 15, 1996; the Clinton Nuclear Generating Station near Clinton, Illinois, on September 5, 1996; The Buchanan, New York, Plant on February 15, 2000; and the Davis-Besse

Nuclear Plant in Oak Harbor, Ohio, on February 16, 2002.[1] From 2002 to the publication of this book, there have been no additional incidents caused by human error when dealing with new equipment at nuclear facilities that the author could obtain.

In all, that is eighteen serious incidents caused by human error involving unfamiliarity with new equipment. During that same period (1955–2021), there were ten incidents caused by human error that did not involve new equipment.[2] Typically, these involved insufficient training, poor management, and simple misjudgment.

THE PROBLEM AND RESEARCH QUESTION

It is clear that the incidence of US reactor core meltdowns being thirty-two times higher than what theory would predict is because additional unaccounted-for factors are violating the base assumptions of the model. The history of human error and nuclear reactor incidents clearly suggests that such an assumption is unfamiliarity with new equipment.

Human error will always be a factor, but it is the goal of safety procedures to avoid or compensate for those errors. Here, I investigate incidents at US nuclear facilities that occurred due to human error by people who interacted with the complex systems involved with a nuclear plant, and especially with its changing technologies. Realizing that the top-level goal is for the safe and efficient operation of nuclear facilities (and the potential application of other complex systems) and that an optimized human-machine system is the desired end state, I focus on the human performance component of this complex system.

Faced with the data, we must ask ourselves whether the severity, cost (in the hundreds of millions, as will be discussed in Chapter 5), and other consequences caused by human errors can be reduced through effectively and efficiently implementing necessary technological changes and human performance activities that directly impact how operations occur at complex nuclear facilities.

The urgency and importance of safely implementing upgrades in equipment are clear when we consider the consequences of nuclear incidents on the facilities involved, the environment, human beings, and the nuclear industry. Several have resulted in radiation being released to the environment and in fatalities. Much research has been conducted on the causes and consequences of adverse incidents at nuclear plants, but not specifically on my area of focus: how technological changes could be a key factor increasing the risk of human error.

Because there are so many outcomes of nuclear incidents, I define their severity in terms of the cost of incidents caused by human error as a result of technological advances, rather than trying to examine other physical and environmental effects associated with the incidents, such as radiation releases, on which most research on nuclear accidents focuses [1]. Put simply, my hypothesis is that technological changes that affect how operators interact within the systems of nuclear facilities exacerbate the cost of incidents caused by human error. My null hypothesis is that technological changes that affect how operators interact within the systems of the nuclear facilities do not exacerbate the cost of incidents caused by human error.

If my null hypothesis is statistically unlikely, then improving the training and procedures associated with facility technology changes could decrease the probability and severity of future incidents. This could result in significant savings both economically and in avoiding adverse impacts on people, the environment, and nuclear industry perception.

My ultimate goal here, however, is not simply to show that new equipment increases the chances and costs of human error. It is to demonstrate a methodology to overcome these challenges by applying the systems engineering process.

To do this, I focus first on the relationship between technological advances and human performance in the complex systems at nuclear facilities to determine whether technological advances in these complex systems increase the cost associated with incidents caused by human error. This is an important question because most nuclear facilities continually update their technology. This continuous technological improvement can create a situation in which the operators must change their routine and method of interacting with the complex system and the new technology.

RESEARCH PURPOSE AND DESIGN

Extensive research exists on the safety procedures used by nuclear facilities, including publications by the International Atomic Energy Agency (IAEA), US Nuclear Regulatory Commission (NRC), and US Department of Energy (DOE) along with independent scholarly research. Also, considerable research has investigated the effect of new technologies on human performance, especially when complex systems are involved. This literature includes the foundational works *Human Error*, by James Reason; *Behind Human Error*, by David Woods, Sidney Dekker, Richard Cook, Leila Johannesen, and Nadine Sarter; and *The Field Guide to Human Error*, by Sidney Dekker. The systems involved with nuclear facilities are some of the most complex systems ever developed. This fact suggests that these systems could be especially exposed to the threat of human error following the implementation of technological changes. I want to understand the relationship between human performance, technological advances, and the complex systems involved with nuclear facilities.

My next step involves an extensive examination of the impact of human factors (including human error) on plant operation and the systems engineering process. Then I can create a model to cultivate human performance-enhanced system design and operation via operator involvement in all stages of the systems engineering process, iterative procedure, and operator training development throughout all stages of the systems engineering process, and early selection and cultivation of suitable operators chosen and groomed specifically for the systems being designed. Using a methodology to optimize the balance between the human and machine sides of the human-machine interface in a system, I can finally suggest a path forward through the systems engineering process. In short, I have a method of incorporating human performance characteristics into the system design and development stages of the systems engineering process. Specifically, I propose the incorporation of human performance attributes into system operational requirements, Technical Performance Measures (TPM), and system testing, evaluation, and validation.

NOTES

1 **May 15, 1996: Morris, Illinois** The Dresden Generating Station, located near Morris, Illinois, is the first privately financed nuclear power plant in the United States. It is owned and operated by Exelon Generation, LLC. There have been three boiling water reactor units within the facility. Unit one became operational in 1960 and was decommissioned in 1978. Unit two was made operational in 1970 and is licensed to operate until 2029. Unit three was commissioned in 1971 and has a license to operate until 2031. Both operating reactors have a maximum capacity of 867 MW. The facility provides power for Chicago and roughly one-quarter of the state of Illinois. It generates enough electricity for approximately 1 million homes [7]. This facility has had a problematic operating history [5]. From 1970 to 1996, it accumulated fines of more than $1.5 million. In May 1996, the water levels surrounding the reactor core dropped to an unacceptably low level, forcing a temporary shutdown of the facility. Frequent changes were made at the facility prior to 1996 in response to the numerous NRC sanctions. It is likely that multiple changes in the technology prior to the 1996 incident contributed to the operators' inability to determine that the water levels were too low [3].

September 5, 1996: Clinton, Illinois The Clinton Nuclear Generating Station located near Clinton, Illinois [8], was commissioned in 1987 and has a license to operate until 2026. The cost of the facility was over $2.6 billion. It is operated by the Exelon Corporation and has a second-generation boiling water reactor produced by General Electric. The operational reactor has a capacity of 1,043 MW. The original owner of the plant was Illinois Power. In 1996, Illinois Power shut down the facility because a reactor recirculation pump failed [7]. Although it was not generally made public at the time, Illinois Power suspected that many of the problems were due to operators' unfamiliarity with how to run the facility and their general inability to use and understand the readings provided by the recently updated user interface. In fact, following the incident in 1996, Illinois Power sold the facility to Exelon Corporation at a substantial loss. The estimated loss for the temporary shutdown in 1996 was $36 million, the original construction cost of the facility exceeded $2.6 billion, and the facility was sold to Exelon for only $40 million [2].

February 15, 2000: Buchanan, New York The Buchanan Plant experienced a shutdown in 2000. It appears that the operators at the Indian Point Energy Center failed to read the system feedback properly and that a Freon leak was allowed to continue for a significant amount of time before it caused the ventilation train chiller to trip [2]. Furthermore, the plant had recently undergone changes to its computer system in anticipation of the new millennium, probably contributing to the operators' failure to fully understand the readings [7]. This can be considered an example of technological advances negatively affecting operators' interaction with the system.

February 16, 2002: Oak Harbor, Ohio In March 2002, the facility maintenance workers at the Davis-Besse Nuclear Plant discovered a football-sized hole located near the reactor vessel head [7]. No adverse incident resulted from this hole, but the NRC required the plant to close for two years while FirstEnergy completed the necessary maintenance. The corrosion of the reactor head was attributable to boric acid. Abundant readings should have caused the operators to suspect that something was wrong. However, the facility had recently undergone numerous technological enhancements to the user interface in its main control center. When questioned, the operators admitted that they did not feel fully confident with the new equipment [3]. Accordingly, human error related to technological changes was definitely a factor in this incident.

2 Incidents: July 24, 1964: Charlestown, Rhode Island; July 16, 1971: Cordova, Illinois; March 22, 1975: Athens, Alabama; November 22, 1980: San Clemente, California; March 20, 1982: Scriba, New York; February 12, 1983: Forked River, New Jersey; September 15, 1984: Athens, Alabama; March 31, 1987: Delta, Pennsylvania; December 25, 1993: Newport, Michigan; August 4, 2005: Buchanan, New York.

REFERENCES

1. Republished with permission of American Society of Mechanical Engineers, from Corrado, Jonathan, and Ronald Sega. 2020. "Impact of Advancing Technology on Nuclear Facility Operation." *ASCE-ASME Journal of Risk and Uncertainty in Engineering Systems, Part B: Mechanical Engineering* 6(1). doi:10.1115/1.4044784; permission conveyed through Copyright Clearance Center, Inc.

2. Perrow, Charles. 1999. *Normal Accidents: Living with High-Risk Technologies.* Princeton, NJ: Princeton University Press.
3. Krivit, Steven B., Jay H. Lehr, and Thomas B. Kingery. 2011. *Nuclear Energy Encyclopedia.* Hoboken, NJ: John Wiley & Sons.
4. Rogers, Mary Jo. 2013. *Nuclear Energy Leadership: Lessons Learned from U.S. Operators.* Tulsa, OK: PennWell Corp.
5. Harrison, R. M., and R. E. Hester, eds. 2011. *Nuclear Power and the Environment.* Cambridge: RSC Publishing.
6. Marques, J. G. 2011. "Safety of Nuclear Fission Reactors: Learning from Accidents." In *Nuclear Energy Encyclopedia Science, Technology, and Applications*, edited by S. B. Krivit, J. H. Lehr, and T. B. Kingery, 127–49. Hoboken, NJ: John Wiley & Sons.
7. Sovacool, Benjamin K. 2011. *Contesting the Future of Nuclear Power.* Hackensack, NJ: World Scientific.
8. Smith, G. 2012. *Nuclear Roulette: The Truth about the Most Dangerous Energy Source on Earth.* White River Junction, VT: Chelsea Green Publishing.
9. Salvendy, Gavriel. 2012. *Handbook of Human Factors and Ergonomics.* 4th ed. Hoboken, NJ: John Wiley & Sons.
10. Sehgal, Bal Raj., ed. 2012. *Nuclear Safety in Light Water Reactors: Severe Accident Phenomenology.* Waltham, MA: Elsevier/Academic Press.

2

Nuclear Incident Severity Determination and Public and Industry Perception of Incidents and the Impact This Has on Nuclear Power's Future

SYSTEMS ENGINEERING AND COMPLEX TECHNOLOGIES

Systems engineering, which is concerned with how to design, integrate, and manage complex systems from the inception and end of their life cycles, has been recognized for over sixty years as essential to the proper development of

DOI: 10.1201/9781003346265-2

complicated systems. It has been applied to a wide range of technological projects such as automobiles, urban infrastructure, environmental controls, aircraft, software, hardware, and ships. Systems engineers are often the technical leaders for vast and complex projects.

Systems engineers rely on the application of relationships and system science to analyze and determine system performance of a product under development. Twenty-first-century systems engineering guides the development of each part within the system through learned heuristics. Both engineering managers and systems engineers understand that the practice of systems engineering has significant value. For this reason, systems engineering concepts and practices are used in nearly all complex projects. Despite the acknowledged importance of systems engineering, some observers argue that it is less fully understood than other engineering disciplines [1].

Because of continually evolving technology and accompanying increases in the level of system complexity, twenty-first-century systems engineering is frequently confronted with greater depths of contextual embedding. Systemic behaviors today present an increasing number of specifications and environmental parameters for consideration [2].

One clear example of a complex system that involves systems engineering, human performance, and technological advances is a nuclear facility. More specifically, this means nuclear power plants, nuclear material processing and fabrication facilities, and enrichment plants. Nuclear facilities are among the most complex systems ever designed and use state-of-the-art technology. They must continually update their systems to remain safe, regulatory compliant, and economically sound. Unfortunately, like any complex system, nuclear facilities are not immune to failure, particularly with regard to human performance. As discussed in Chapter 1, many of the adverse incidents that have occurred at nuclear facilities have been the result of some type of human error [2]. Regulations and standards have, as they must, been established to mitigate the risk of human error.

INTERNATIONAL ATOMIC ENERGY AGENCY AND US NUCLEAR REGULATORY COMMISSION AND THEIR ROLE IN REGULATORY AND SAFETY MATTERS

The International Atomic Energy Agency (IAEA) is a crucial stakeholder in the global nuclear industry. It was established in 1957 and still operates as an autonomous organization promoting the use of nuclear power for peaceful purposes. The IAEA also seeks to reduce the use of atomic energy for weapons or other military purposes. It functions independently from the United Nations but reports to the UN General Assembly [3].

The IAEA has three basic concerns: safeguards and verification, science and technology, and safety and security [4]. These three concepts underlie all missions carried out by the IAEA. In its interaction with the UN, the IAEA generally interfaces with the Security Council. The IAEA is composed of three general bodies: the Secretariat, the General Conference, and the Board of Governors [5].

The IAEA has three primary functions. It acts as a hub for the myriad fields of science, and it considers how nuclear technology can be peacefully applied. It also ensures the security and safety of atomic facilities through standards and by providing information on the nuclear industry. To fulfill its mission, the IAEA inspects the world's nuclear facilities to ensure that they are being run properly and used in a peaceful manner. As an illustration of the diversity of nuclear science activities, in 2004 the IAEA introduced the Program of Action for Cancer Therapy (PACT), in a response to developing countries' need for modern treatment programs using radiotherapy [6].

Although nearly all countries with a nuclear industry are regulated by the IAEA, the United States is an exception, monitoring its nuclear facilities through the independent government agency, the US Nuclear Regulatory Commission (NRC), formed following the Energy Reorganization Act of 1974 as the successor agency to the US Atomic Energy Commission (AEC). The NRC licenses and regulates the nation's civilian use of radioactive materials to protect public health and safety, promote the common defense and security, and protect the environment. The NRC's regulatory mission covers three main areas: commercial reactors for generating electric power, and research and test reactors used for research, testing, and training; uses of nuclear materials in medical, industrial, and academic settings and facilities that produce nuclear fuel; and the transportation, storage, and disposal of nuclear materials and waste, and decommissioning of nuclear facilities from service.

Nationally, the NRC oversees more than a hundred nuclear reactors that produce power along with several fuel cycle and waste facilities. There are also thirty-three nuclear reactors that have been permanently shut down and several new reactors under construction.

There are multiple levels of regulatory oversight of US nuclear facilities. The first of these consists of resident inspectors who are charged with monitoring the plant's daily operations. Resident inspectors are generally found at nuclear power plants, whereas fuel cycle facilities are assigned project inspectors that are responsible for the plant, but not permanently stationed on site. NRC inspection teams inspect all aspects of plant operation and administration throughout an inspection cycle. Special inspection teams can also be chartered to investigate events, violations, and possible whistleblower reports.

The United States currently has cooperation agreements with the IAEA. Prior to the early 1970s, the United States had no agreement with the IAEA, causing concerns that the US would have an industrial and commercial advantage over other countries when using nuclear energy for peaceful purposes. In response to this concern, the United States entered into an agreement with the IAEA for nuclear plant inspection. The agreement excludes any facilities that are producing or using nuclear power for national security. In 1993, the United States agreed to place any nuclear material in excess of its defense needs in storage according to IAEA standards. The United States also uses the International Nuclear and Radiological Event Scale (INES), which the IAEA introduced in 1990.[1]

THE INTERNATIONAL NUCLEAR AND RADIOLOGICAL EVENT SCALE

The INES provides simple and readily understandable information about nuclear incidents. The scale is logarithmic and similar in concept to the scale that measures the magnitude of earthquakes; each new level on the scale is ten times as severe as the previous one. For earthquakes, intensity can be evaluated in a quantitative fashion. However, judgments of the severity of a nuclear incident are more subjective and require extensive investigation. For this reason, an INES level is generally not assigned to an incident until a significant period of time following the event. Unfortunately, this means that the scale is sometimes not useful for rapid deployment of disaster aid.

The INES has seven levels, from the least severe to the most problematic: anomaly, incident, serious incident, accident with local consequences, accident with wider consequences, serious accident, and major accident. The first three levels are sometimes grouped together under the category of atomic incidents; the highest four categories are referred to as nuclear accidents. An eighth level, referred to as a deviation or Level 0, indicates an event with no safety significance. For example, a reactor might need to be shut down because a cooling circuit leaked, but the event is not an atomic incident or accident if it does not result in the release of radioactive substances [7]. See Figure 2.1 for a graphic of the INES.

The first level of the scale, an "anomaly," is achieved when a member of the public is exposed to radiation exceeding the yearly statutory limits. This

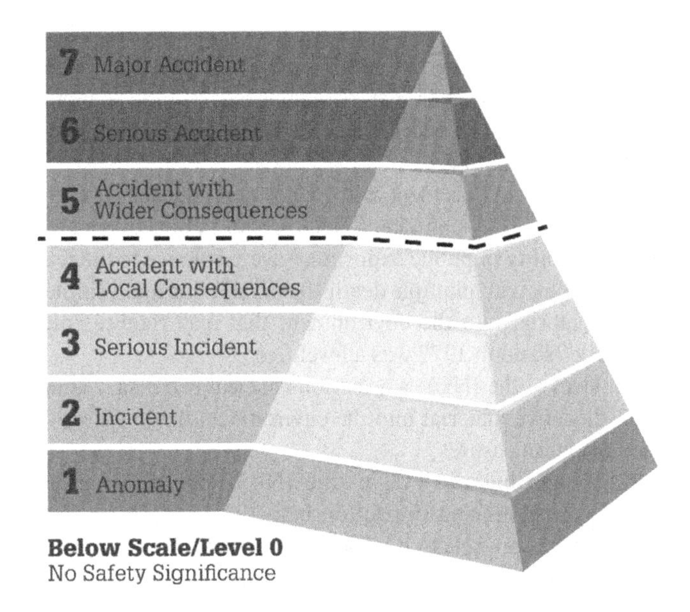

FIGURE 2.1 The International Nuclear and Radiological Event Scale (From US Nuclear Regulatory Commission at www.nrc.gov).

could be a minor problem involving safety components. However, there are significant defenses against harm. The second level, an "incident," results in a worker's being exposed to radiation beyond the yearly statutory limits. It may also involve a member of the public being exposed to a radiation level in excess of 10 millisieverts (mSv). For context, a dose of 1,000 mSv causes radiation sickness, which can include nausea and hemorrhaging, but is not lethal. A single dose of 5 mSv would kill roughly half of the exposed individuals within a few weeks. Typical equipment problems at the 10 MSv-level include improperly packaging of radioactive sealed sources, the locating of an orphan source of radioactivity that is still sealed, or the discovery of a significant problem with safety provisions. The impact of Level 2 problems on radiological barriers and controls can result in substantial contamination at the facility. Radiation levels in the operating area will exceed 50 mSv.[2]

Level 3 is considered a "serious incident." Although there may be severe contamination within the problem area, it is not likely that the public would be exposed to significant radiation. Workers in the area may be exposed to radiation levels exceeding ten times the yearly limit. People may experience non-lethal health problems such as radiation burns. An example of a Level 3 incident is the delivery of a sealed, highly radioactive source without the observance of adequate procedures and standards.

The fourth level of the INES is an "accident with local consequences." Substantial amounts of radioactive material may be released, and public exposure is highly probable. The fuel may be damaged, resulting in more than 0.1 percent of the core inventory being released. The results will likely include at least one death from radiation.

The fifth level is an "accident with wider consequences." The probable result here is the release of significant quantities of radioactive material, with a substantial probability of public exposure. Severe damage to the reactor core may also occur, along with multiple deaths from radiation and a limited release of radioactive material into the environment that may require countermeasures. Three Mile Island in 1979 was a level five.

The sixth level of the INES is a "serious nuclear accident." This includes a release of radioactive material into the environment that requires the use of well-planned countermeasures.

The seventh and highest level of the INES is a "major nuclear accident," which has a significant impact on both the environment and people. Radioactive materials are released in large quantities, and both extended and well-planned countermeasures must be used. Only two events in history have reached this level: Chernobyl and Fukushima.

The IAEA is responsible for regular safety analyses of nuclear facilities. These analyses evaluate the physical environment at the nuclear plants. They are intended to demonstrate that the proper safety requirements have been met for all types of initiating events. These requirements include creating policies for ensuring the integrity of the barriers preventing the release of radioactive materials.[3]

PROBABILISTIC AND DETERMINISTIC SAFETY ANALYSIS

The IAEA conducts two basic types of analyses of nuclear plants: probabilistic and deterministic safety analysis. Deterministic safety analysis involves predicting the responses to possible initiating events [8]. There is a precise set of rules and criteria for acceptance. Usually, the criteria and rules focus on structural, thermo-mechanical, radiological, thermo-hydraulic, and neutronic aspects of the plant. A variety of computational tools are used to do these analyses. The computations are done for predetermined modes of operation and specific states of the systems. The events include severe accidents with core degradation beyond the design basis, postulated accidents, and accidents that are anticipated and transient. The resulting computations yield physical variables and time related as well as spatial dependencies. These dependencies can include concentrations of radionuclides, chemical composition, tangible impacts and stresses to structural materials, and, in the case of nuclear power plants, problems with coolant flow rates, temperature, pressure, thermal power, and neutron flux [9]. When an assessment of prospective radiological consequences is conducted, the dependency is the potential dose received by the public or plant workers.

When a deterministic safety analysis is prepared for the purposes of plant design, it is characterized by bounding analysis and conservative assumptions [9]. This can be done using an iterative process within the design phase of the project. The limiting case for the minimum margin of the acceptance criteria is determined for each of the several postulated sequences or initiating events. To understand the specific limiting case for a given transient or set of transients, there must be consideration of the consequential failures that have resulted from the external or internal initiating event [10].

A sufficient set of best estimates for conservative assumptions for the boundary or initial conditions must be used [9]. Furthermore, independent failures should be addressed that are both coincident and limited in number. Operator error must also be included. The frequency of accident occurrences will decrease as each of the coincident independent failures is considered (i.e., because some failures occur together, the overall chance of a failure is less than the sum of the independent failures). Only the combinations of the transients with a frequency within the design basis should be of concern [10].

The time frame used for a scenario should encompass everything that occurs up to the moment when a plant achieves a stable and safe state of operation [8]. The states must be defined. In the case of nuclear reactors, it is assumed that a stable and safe state is present when the reactor core is properly covered and long-term heat removal has been achieved. In addition, the core must be sub-critical (i.e., no fission occurring and a reasonable shutdown) and include a given margin. The safety analysis should include provisions for removing the fuel in a secure manner and storing it in another location after it has cooled [11].

To ensure a proper defense in-depth, all the barriers and credible mechanisms of failure must be considered. Limiting faults such as rod ejections and secondary breaks in nuclear power plants should be included in the deterministic safety analysis. The leaks prior to the break criterion from best estimate analysis can be used in the definition of requirements for structures, components, and systems [9].

The second type of analysis, probabilistic safety analysis, is used to ascertain the probability of damage being caused by a failure of each barrier [9]. This type is especially useful for evaluating the risk from low-frequency sequences that can lead to the damaging of a barrier. In contrast, deterministic analysis is better suited for events that occur frequently and are anticipated by the acceptance criteria. Probabilistic safety analysis is used for evaluating whether the in-depth defenses are adequate. This might include events such as a severe loss of coolant in nuclear power plants [8].

Both deterministic and probabilistic safety analyses provide important data on accident scenarios [10]. Whereas deterministic safety analysis is used to identify challenges to the physical barriers or integrity of systems, probabilistic safety analysis uses data and codes for estimation. These data and codes must be consistent with the objectives of the analysis. The results of the probabilistic analysis can be understood as supporting the results of the more conservative deterministic analysis [9].

When a probabilistic safety analysis is performed, a fault tree is often used to confirm whether the assumptions made in the deterministic calculation about the availability of the systems are correct [9]. For example, this approach might be applied in determining the potential for common cause failures or the establishment of minimum system requirements. The probabilistic safety analysis fault tree can also be used to determine whether the technical specifications are adequate and to identify the most important individual potential failures [11].

NUCLEAR POWER AND PUBLIC OPINION

As with Three Mile Island, the March 2011 disaster at the Fukushima power plant had a significant impact on public opinion concerning atomic policies, though that impact seems to have been much greater. For example, the Chinese government stopped all its nuclear projects. Public support for any type of atomic power in the Republic of Korea disappeared. Germany and Belgium enacted legislation to eliminate nuclear power by the second decade of the twenty-first century. Switzerland and the Netherlands also stopped any projects to build additional atomic power plants. Other world governments also revisited their plans for nuclear power [12].

The global reaction to the Fukushima nuclear power plant accident also meant that in 2011 nineteen atomic reactors were completely shut down, eighteen of them specifically as a consequence of the problems at Fukushima. As of the publication of this book, only seven of these reactors have resumed operation since that time. Germany, the first country to begin operating a new reactor following the Chernobyl incident, closed eight reactors after Fukushima. Fourteen months after that nuclear accident, Japan had only one reactor still in operation [4].

In 1992, the World Nuclear Industry Status Report was established to assess the global impact of the Chernobyl incident on the nuclear industry [13]. The report predicted that fewer atomic plants would be constructed. During the first decade of the twenty-first century, this prediction has been confirmed by the rapid growth of competitors to nuclear power such as solar and wind power [14].

Public concerns regarding safety are not the only disadvantages facing atomic power plants, as their construction frequently involves cost overruns, construction delays, and long lead times. Nuclear power is increasingly viewed as a type of risky investment that many countries are choosing to avoid [15]. For instance, France shut down its oldest nuclear plant in 2020 after 43 years of operation, the first in a series of closures the government has proposed as part of a broad energy strategy to rely more on renewable energy sources. That strategy would see French dependence on nuclear energy from supplying three-quarters of its electricity to about half by 2035.

In 2012, thirty-one countries had nuclear fission reactors for the purposes of producing energy [4]. The only new reactor to come online between 2010 and 2012 was the Bushehr reactor in the Islamic Republic of Iran. In 2011, more than 2,500 kilowatt hours of electricity were produced at nuclear power plants, approximately equivalent to the amount produced in 2001 and 5 percent below 2006, which was the highest production year for nuclear power. In 1993, the proportion of all electricity produced by atomic power reached its highest level at 17 percent; by 2011, the percentage had fallen to 11 percent [16].

In July 2012, there were 429 nuclear reactors operating in thirty-one countries [12]. This was a decline from the 444 plants operating in 2002. By July 2012, Japan had shut down forty-nine of its fifty nuclear power plants. Over the course of 2012, thirteen countries were building new nuclear power plants, down slightly from the fifteen countries reported in 2011. In 2012, the total number of nuclear reactors under construction increased to fifty-nine. However, this was far below the peak of 234 nuclear construction projects in 1979 [17].

Even though reactors continue to be built, their dwindling numbers illustrate the uncertainty of the use of nuclear power [4]. As of 2012, nine reactors had been under construction for over twenty years. The longest construction period belongs to Tennessee's Watts Bar Nuclear Generating Station Unit 2, which began construction in 1973 but, due to several complications, was not completed until 2015 [16]. Currently, eighteen nuclear power plants under construction have been associated with significant construction delays. More than 70 percent of the plants under construction are located in Russia, China, and India. These three countries have not provided reliable information about the status of their atomic power plant construction. Nevertheless, it is generally accepted that more than half of the Russian nuclear power plants under construction are experiencing delays of at least several years [18].

Ordinarily, long lead times for atomic power plants result from long-term planning, extended construction times, and lengthy licensing procedures. These projects also require extensive site preparation and complex financing. All these obstacles have reduced the number of new nuclear power plant construction projects or grid connections. The average operating age of the atomic power plants has been steadily increasing and is now at about twenty-seven years. Some of the facilities have been operating from forty to sixty years [12].

The US licenses nuclear power plants to operate for forty years, but many countries do not place a time limit on their licenses. France, which first began operating an atomic reactor with pressurized water in 1977, has a policy of conducting an in-depth inspection of nuclear power plants once each decade and has permitted only two plants to operate beyond thirty years. One of those plants has been permanently shut down and the other is scheduled for permanent closure in the near term before they reach their forty-year anniversary of operation.

While numerous costly incidents have occurred at US nuclear facilities, similar events have occurred worldwide and are not unique to any one particular country. In nearly every incident, the cause has been isolated and steps have been taken to alleviate problems and prevent future incidents. However, thousands of upgrades to various pieces of equipment take place over the lifetime of a nuclear facility. The vast majority of these changes do not result in difficulties. Although there is no way to know with certainty what problems have been averted through technological improvements, it seems reasonable to assume that these upgrades have reduced the number of incidents; in other words, fewer problems have been created by the new technology than would have resulted without the changes [19]. However, it is also possible that the introduction of new technology increases the risk of human error. Since changing the technology associated with nuclear facilities is an ongoing and necessary process, the question of whether technology in the complex systems of nuclear facilities decreases human performance and increases the chance of incidents is an important question to investigate.

NOTES

1 For more, see www.nrc.gov and www.iaea.org.
2 For more, see Lars Högberg, "Root Causes and Impacts of Severe Accidents at Large Nuclear Power Plants," *AMBIO* 42, no. 3 (2013): 267–84, http://doi.org/10.1007/s13280-013-0382-x.
3 For more, see www.iaea.org.

REFERENCES

1. Blanchard, Benjamin, and Wolter Fabrycky. 2011. *Systems Engineering and Analysis*. 5th ed. Upper Saddle River, NJ: Pearson Education.
2. Marques, J. G. 2011. "Safety of Nuclear Fission Reactors: Learning from Accidents." In *Nuclear Energy Encyclopedia Science, Technology, and Applications*, edited by S. B. Krivit, J. H. Lehr, and T. B. Kingery, 127–49. Hoboken, NJ: John Wiley & Sons.
3. Mengolini, Anna, and Luigi Debarberis. 2012. "Lessons Learnt from a Crisis Event: How to Foster a Sound Safety Culture." *Safety Science* 50(6): 1415–21. doi:10.1016/j.ssci.2010.02.022.
4. Perko, Tanja, Catrinel Turcanu, and Benny Carlé. 2012. "Media Reporting of Nuclear Emergencies: The Effects of Transparent Communication in a Minor Nuclear Event." *Journal of Contingencies & Crisis Management* 20(1): 52–63. doi:10.1111/j.1468-5973.2012.00663.x.
5. Budnitz, Robert J. 2010. "Status Report on the Safety of Operating US Nuclear Power Plants (Why Experts Believe That Today's Operating Nuclear Power Reactors Are Much Safer Than They Were 20 Years Ago)." Chap. 3 in *Nuclear Power and Energy Security*, edited by Samuel A. Apikyan, and David J. Diamond, 109–19. Netherlands: Springer. doi:10.1007/978-90-481-3504-2_15.
6. Högberg, Lars. 2013. "Root Causes and Impacts of Severe Accidents at Large Nuclear Power Plants." *AMBIO* 42(3): 267–84. doi:10.1007/s13280-013-0382-x.
7. Marques, J. G. 2011. "Safety of Nuclear Fission Reactors: Learning from Accidents." In *Nuclear Energy Encyclopedia Science, Technology, and Applications*, edited by S. B. Krivit, J. H. Lehr, and T. B. Kingery, 127–49. Hoboken, NJ: John Wiley & Sons.
8. Sugiyama, G., J. S. Nasstrom, B. Probanz, K. T. Foster, M. Simpson, P. Vogt, and S. Homann. 2012. *NARAC Modeling during the Response to the Fukushima Daiichi Nuclear Power Plant Emergency* (No. LLNL-CONF-529471). Livermore, CA: Lawrence Livermore National Laboratory.
9. Hashemian, H. M. 2010. "Aging Management of Instrumentation and Control Sensors in Nuclear Power Plants." *Nuclear Engineering and Design* 240(11): 3781–90. doi:10.1016/j.nucengdes.2010.08.014.
10. Perkins, Richard H., Michelle T. Bensi, Jacob Philip, and Selim Sancaktar. 2011. *Screening Analysis Report for the Proposed Generic Issue on Flooding of Nuclear Power Plant Sites Following Upstream Dam Failures*. Washington, DC: Division of Risk Analysis and Applications, Office of Nuclear Regulatory Research, U.S. Nuclear Regulatory Commission.
11. Prasad, R., L. F. Hibler, A. M. Coleman, and D. L. Ward. 2011. *Design-Basis Flood Estimation for Site Characterization at Nuclear Power Plants in the United States of America* (No. PNNL-20091; NUREG/CR-7046). Richland, WA: Pacific Northwest National Laboratory.
12. Sharma, Rakesh Kumar, and Rajesh Arora. 2011. "Fukushima, Japan: An Apocalypse in the Making?" *Journal of Pharmacy and Bioallied Sciences* 3(2): 315–27. doi:10.4103/0975-7406.80756.

13. Saey, Paul R. J., Matthias Auer, Andreas Becker, Emmy Hoffmann, Mika Nikkinen, Anders Ringbom, Rick Tinker, Clemens Schlosser, and Michel Sonck. 2010. "The Influence on the Radioxenon Background during the Temporary Suspension of Operations of Three Major Medical Isotope Production Facilities in the Northern Hemisphere and during the Start-Up of Another Facility in the Southern Hemisphere." *Journal of Environmental Radioactivity* 101(9): 730–38. doi:10.1016/j.jenvrad.2010.04.016.
14. Lelieveld, J., D. Kunkel, and M. G. Lawrence. 2012. "Global Risk of Radioactive Fallout after Major Nuclear Reactor Accidents." *Atmospheric Chemistry and Physics* 12(9): 4245–58. doi:10.5194/acpd-12-19303-2012.
15. Thomson, Elspeth. 2011. "China's Nuclear Energy in Light of the Disaster in Japan." *Eurasian Geography and Economics* 52(4): 464–82. doi:10.2747/1539-7216.52.4.464.
16. Christoudias, Theodoros, and Jos Lelieveld. 2013. "Modelling the Global Atmospheric Transport and Deposition of Radionuclides from the Fukushima Dai-Ichi Nuclear Accident." *Atmospheric Chemistry and Physics* 13(3): 1425–38. doi:10.5194/acpd-12-24531-2012.
17. Shultz, James M., David Forbes, David Wald, Fiona Kelly, Helena M. Solo-Gabriele, Alexa Rosen, Zelde Espinel, Andrew McLean, Oscar Bernal, and Yuval Neria. 2013. "Trauma Signature Analysis of the Great East Japan Disaster: Guidance for Psychological Consequences." *Disaster Medicine and Public Health Preparedness* 7(2): 201–14. doi:10.1017/dmp.2013.21.
18. Gang-yang, Zheng, Song Yan, and Zhang Zhi-jian. 2011. "Research on Public's Perception of Risk Propagation Process of Nuclear Power Plant." Paper presented at the International Conference on Information Systems for Crisis Response and Management (ISCRAM), Harbin, China, November 2011. doi:10.1109/ISCRAM.2011.6184041.
19. Woods, David D., Sidney Dekker, Richard Cook, Leila Johannesen, and Nadine Sarter. 2010. *Behind Human Error*. Farnham: Ashgate.

Why Nuclear Incidents Happen 3

A Detailed Review of Human Performance, Human Errors, and Technology Interface

HUMAN PERFORMANCE AND TECHNOLOGY COMPLEXITY

Each time significant technological developments offer the promise of greatly assisting people's lives, failed systems and prototypes inevitably result to some degree. When researchers examine the effects of changes in technology, they often discover unintended and unanticipated consequences. Individuals using the new technologies frequently make performance errors because they must adapt to increasingly complex technology. Rather than assisting the user, these new technologies can add burdens, which are especially problematic during crucial phases of complex tasks [1].

The pattern of human performance degradation when novel technologies are introduced occurs in a wide range of endeavors. For example, the implementation of new systems of airplane cockpit automation can be associated with a decline in pilots' performance; their reaction times and number of

DOI: 10.1201/9781003346265-3

errors may increase. The same is true in virtually all industries, including the nuclear industry, where human errors that affect system operation are unacceptable [2].

Although considerable research has been conducted on the human-machine interface, a wide range of problems still exists. There remains a conflict between the optimism of technology developers and the real-life operational difficulties that accompany the introduction of these systems. The developers nearly always claim that the new technology will result in performance improvements. However, due to the operational complexities introduced, the technology may actually decrease the performance of those interacting with the system. Unfortunately, the complexities confronting operators are difficult for design teams to predict. To understand the complexity surrounding human interaction with advancing technologies, the human performance model must be examined and the concepts of escalation, active error, latent error, foreseeable error, and unexpected events should be examined [2].

To fully understand how technological advancements interact with human performance at nuclear plants, we should consider the principle of escalation [3] or the idea that problems tend to increase. What begins as a small problem or error leads to an increase in the coordinating and cognitive demands required to accomplish a task. This frequently results in larger errors and an increase in problems. In general, there is a positive relationship between the scope of a problem and the amount of information processing necessary to cope with it. When there are more problems in the underlying system, the additional information processing needed to resolve the situation increases. A more complex system requires greater effort to deal with unexpected problems or errors. As technology progresses, the complexity of the systems involved grows as well [4].

MODELING HUMAN PERFORMANCE

Understanding human errors increases through modeling human performance. This section will explain one such model. The reader should recognize that the explanations given for the behaviors represent an ideal situation. Error-likely situations occur when the individual involved has deviated from the behaviors described by the model.

Although many heavily automated technical systems exist today, all of them rely on routine human interaction as an integral characteristic of normal system functioning [5]. Operators must ensure that the proper conditions are present for the system to operate normally and must intervene when abnormal conditions exist so as to restore the system to a safe configuration. They must also account for any unforeseen problems with the system or compensate for anything that has been structured inappropriately due to design flaws. Many of these automated systems play a vital role in society, and tragedy can result when they are not supported properly. For this reason, increased attention has been devoted to human performance and human error when interacting with systems [6].

We need reliable models that ensure the maintenance of a high level of human performance when people are interacting with complicated and automated systems [5]. This requires the understanding of different kinds of error. Quantitative models have been used to do performance analysis and system design in vehicle control for some time. Attempts have been made to extend the models used for vehicle control to other types of human decision-making. One such attempt is optimal control theory. The first of the vehicle control systems to be analyzed was in the field of aviation [7].

The optimal control model is not necessary if activities by people are no longer included in the control task [8]. In these cases, the concern is with an overall interface manipulation skill. Decision models can be constructed in these cases through independent development and a direct approach. Instead of a single quantitative model for predicting human performance, which would account for nearly any situation, it is likely that a set of models will be more applicable and reliable. Each of these models can be applied to particular work conditions and combined with a qualitative framework that will define and describe the relationships [6].

When seeking to understand human performance while interacting with a complex system, we must remember that people are not merely deterministic devices engaged in input and output [7]. Instead, they are often goal-oriented and will pursue information that they consider relevant to achieve their objectives. People behave in a teleological fashion, meaning their behavior is frequently modified as they seek to achieve their goal. Furthermore, this behavior sometimes does not depend on feedback received while the person is engaged in the activity. The factor of experience during previous attempts can also have an impact. People engage in reasoned reflection and will frequently control behavior systems through selection. In this case, the selection is represented by human design choice when interacting with a complex system [9].

Human movement and position within the physical environment are thus not controlled by a simple feedback loop. People adapt to unfamiliar situations based on their previous experiences of successful patterns of behavior. This process overcomes the limits of the human sensory system with regard to immediate feedback. In other words, the humans interacting with the system respond too quickly for them to learn while they are interacting. Instead, people rely on their memory of previous attempts to interact with a complicated system [5].

When human beings rely on a higher level of conscious planning, they usually engage in a complex series of activities as well as feedback correction while working on a task. Changes to their behavior are caused by mismatches between the outcomes and goals. This is generally an inefficient process when one is working with a complex system functioning at a rapid pace. Therefore, when people are engaged in familiar activities, they will resort to the use of a set of rules that have been successful in the past [8].

Skill-Based Behaviors

Human behavior is beyond complex, but a helpful classification can be made by recognizing the distinctions among skill-based behaviors, rule-based behaviors, and knowledge-based behaviors. Skill-based behaviors involve sensory motor performance in which one engages during activities that follow from a certain intention [7]. These behaviors occur without an individual's conscious control. They represent highly integrated, automated, and smooth patterns of behavior. Only on certain occasions is performance based on some type of feedback control that involves motor output in response to the observation of error signals. In many skilled sensory motor tasks, the human body becomes a type of control system with multiple variables that continuously synchronize movements according to the response of the environment. The performance in these cases includes feed-forward control, a command signal from an operator to a source elsewhere in its external environment and is dependent upon an efficient and flexible internal model of the defined complex system. The feed-forward nature of the control must be assumed to explain how coordinated movements can occur rapidly such as in sports or when one is operating a vehicle. Experiments have demonstrated how the feed-forward control takes place in complicated industrial control tasks [6].

Controlling voluntary movements is an immensely complex process [7]. The success of these rapid movements will be independent of how the limbs were positioned initially. The person will function according to different schemata used to generate the complex movements. The schemata access the individual's dynamic internal map of the environment. Sensory input is not generally involved with these types of movements. In other words, the input from the environment does not realign or update the individual's internal information. Performing complicated tasks such as walking on a straight line or drinking from a glass must be understood as an integrated whole and cannot be broken down into elements [9].

Usually, the performance of skill-based behavior is continuous [5]. Higher levels of control are possible and will take the form of conscious intent to make changes in the skill such as moving faster or more accurately. In some cases, performance will include skilled routines that can be isolated. In these cases, the routines are sequences that guide the process of conscious execution. Many human activities involve sequences of skilled activities, as a response to the specific situation [8].

Rule-Based Behaviors

Meanwhile, rule-based behaviors consist of sequences of subroutines for work situations, which are familiar to the individual and can be controlled through the use of previously established rules [5]. These rules are often derived on an empirical basis, based on the success of previous attempts to engage in the activity. Sometimes they are communicated from other people at the time of construction. They can also be gained from the process of consciously solving a problem or developing a plan [6].

Performance is goal-oriented and structured [9]. The feed-forward control is based on stored rules. Frequently, the goal will not be explicitly understood but, rather, is implicit in the situation and automatically results in the release of the appropriate stored rules. This is a teleological type of control, as the rule has been developed through previous successful experiences. This control will evolve according to the behaviors that work best. The rule reflects functional properties that serve to constrain the behaviors exhibited by the environment. It is usually based on properties discovered through prior empirical investigations [7].

In most cases, a goal will be reached only through a considerable sequence of acts during which direct feedback correction related to the goal is not possible. The feedback correction that occurs during performance of a task requires a functional understanding as well as analysis of the responses provided by the environment. This can be considered a type of independent, concurrent activity that is occurring at a higher level and is knowledge based [7].

The distinction between rule-based and skill-based behaviors is not always clear and can depend on the individual's attention level as well as on their level of training [8]. Most often, skill-based performance occurs without conscious attention. For this reason, the actors will not be able to explain how they control performance or how they have used information to guide their performance. However, rule-based behavior occurs at a higher level and is usually based on some type of explicit knowledge. In this case, the person will usually be able to report on the rules involved [6].

In an unfamiliar situation, the individual may be interacting with an environment for which previous experience has provided no rules by means of which to control the situation. In such a context, the performance must be controlled at a higher level of conceptual understanding through the application of knowledge-based behaviors [5].

Knowledge-Based Behaviors

When a knowledge-based behavior is occurring, the goal has been explicitly formulated [9]. The individual develops a useful plan after careful consideration of multiple options. The plans are tested according to whether they can achieve the goal. These tests can take the form of physical processes, which consist essentially of trial and error. They can also be performed conceptually if the individual understands the functional properties within the environment and can predict accurately the effects that the plan will produce. This type of behavior involves functional reasoning in which the person has a mental model of the system involved [7].

Symbols, Signs, and Signals

The information gained from absorbing the environment is an important part of human performance. The type of information varies according to the category of behavior [8]. Information gained through observing the environment may be perceived in a variety of ways. This is also true for the interface between humans and complex machinery or systems. A major reason for problems with the human-machine interaction is that an unfamiliar situation may cause an individual to misinterpret information while shifting from one motor behavior to another and misread the relevant cues [6].

During skill-based behavior, the perceptual motor system synchronizes the individual's physical activity by operating as a type of continuous control system [7]. The system manipulates external objects and enables the individual's body to navigate within the environment. To accomplish this control, information taken from the environment must be in the form of time and space signals. The signals are a type of quantitative indicator that is continuous and can be applied to the time-based behavior occurring in the environment. The signals do not have meaning or any significance unless they are applied as a type of direct physical data related to time and space. Individual performance occurs on a skill-based level and is released through the features that are assigned to patterns of information due to prior experience. This process replaces individual participation in the environment with feedback from time and space control outcomes. Instead, the information acts as a sign that can initiate action [9].

When an individual is engaging in rule-based behaviors, the information will be primarily perceived as a sign. In this case, the information will modify or activate some predetermined manipulation or action. The signs are a reference to proper behaviors or situations, which are based on prior experiences. They are not a reference to functional properties or concepts related to the environment. The signs are often labeled with names that refer to the situation or to states within the environment. They may also represent the individual's task or goals. The signs may be used only to modify or select, and thereby to control, the sequence of subroutines. They cannot be a part of functional reasoning or involved in the generation of new actions. They also cannot predict possible responses in the environment [5].

For information regarding the behavior occurring in the environment to be useful in relation to causal results, the data must be comprehended as a type of symbol [5]. The signs are references to rules or precepts for action. The symbols include concepts, which are tied to the functional properties and may be used for computation and reasoning through a suitable representation. The signs can be understood as a type of external reference to the actions and states of the environment. The symbols are a reference to the internal representation needed for planning and reasoning. Stated succinctly, rule-based behaviors rely on signs, whereas knowledge-based activities are dependent on symbols [9].

The distinction as to whether perceptual information is a symbol, sign, or signal does not depend on the form of the information, but on the context in which the data have been observed [8]. This will be determined by the expectations and intentions of the individual perceiving the phenomenon. The three levels of behavior are characterized by the use of information in different ways. From the view of information processing, the distinction is clear [9].

The signals are the sensory data that represent variables in time and space according to their configuration within the environment [7]. A person can process this information as a type of continuous variable. The signs represent a state within the environment regarding certain conventions as they apply to acts. The signs have features present within the environment and are associated with connected conditions for the actions. Generally, the signs are not processed directly. They serve merely as a method of activating the stored behavior patterns. The symbols include properties, relations, variables, and other information, which can be processed formally. Symbols consist of abstract concepts, which are defined by and related to a formal structure. This structure is applied to the processes and relations by which conventions are associated with features in the external world [6].

Within the context of humans and machines, information functions as a type of time-space signal [9]. The signals are processed in a direct manner and become part of the dynamic control structure for motor performance. They are separate from the information of signs, which can modify the actions to a higher order of abstraction [8].

Errors and Limitations

Within the domain of knowledge-based actions, the causal and functional properties of the environment may be represented in a variety of ways [6]. A variety of problems can occur at the level of the human data processor when it is interacting with a complex physical environment. The constraint of humans' attention span limits the elements of the problem that can be processed simultaneously to only a few. Therefore, when there exists a complex net of causal relations, the environment must be understood as a type of chain of mental operations. This situation gives rise to phenomena such as the point of no return and law of least resistance. These are strategies that depend on sequences of relatively simple operations and may be preferred intuitively. People often exhibit little tendency to pause within a certain line of reasoning in order to develop parallel paths or alternative explanations [8].

An effective method of overcoming the limitations of our attention span may be to modify the data processing occurring in the mind [6]. The mental model can be altered so that the causal structure better fits the specific task and optimizes a transfer of previous successful results. This method will minimize the requirement for new information. The human cognitive process operates efficiently only when there is an extensive use of the model transformations, combined with simultaneous updating of the mental models. This is true for all categories of inputted information. The type of updating that occurs is generally below the threshold of conscious control or attention [7].

With regard to analyzing verbal protocols, several strategies can be used for model transformation, which facilitates cognitive data processing [7]. One of these is aggregation, which involves taking elements of the representation and placing or aggregating them into chunks or units. Another strategy is abstraction, which involves representing the properties for the environment or a system by transferring them so that they become a category of a higher-level model. The use of ready-made solutions and technologies can also be an effective strategy. This approach involves transferring the representation to a category within a model that has an already evident solution or rules that may be available to generate the solution [6].

An abstraction hierarchy has been formed to analyze the verbal protocols of process plant control and computer maintenance [9]. In this hierarchy, systems' functional properties are represented through concepts that belong to different levels of abstraction. The lowest of these levels represents the physical form of the system, or its material configuration. The next higher level of abstraction is represented through the functions or physical processes of the components in the system. This level is presented in a language associated with particular mechanical, chemical, or electrical properties. Continuing to move upward, the next level of abstraction includes the functional properties that are represented by general concepts. At this level, there is no reference to the physical equipment or processes involved with the functions being implemented [8].

At the lower levels of abstraction, the component configuration for the physical implementation will match the elemental descriptions [6]. At the next level of abstraction, the changes in the properties of the system are represented through removing details regarding the material or physical properties. The information added at the higher abstraction levels governs the functioning of the elements at the lower levels [9]. When a system is manmade, the principles of the higher levels can be derived according to the purpose of the system. For a change in the level of abstraction to occur, there must be a shift in the structure and concepts of the representation, along with a change regarding the information deemed suitable for characterizing the operation or function at the various levels. The observer will ask various questions regarding the environment according to the nature of their internal representation [7].

Important functions within human-machine systems are related to the correction of circumstances that have resulted due to faults or errors [8]. The events are described as faults or errors only in reference to the normal function or intended state of the system. This means that the functional meaning of the system must be predetermined. The model's functioning at the various levels of abstraction can play a role in coping with systems that are plagued by errors [5]. The reasons attached to the proper functions are taken from a top-down approach, beginning with the functional purpose. A relatively clear difference exists between the propagation of faults and causes and the reasons for functions within the hierarchy. The role that the abstraction hierarchy plays is evident in vertical protocols, which are involved in the diagnostic searches of information-processing systems. In these cases, the diagnosticians must consider the functions of the system at a variety of levels. The person will identify the information flow as well as the functional state by approaching the subject from a top-down perspective [9].

Translation for the Operator in the Industrial Setting

An organization of the different types of information processing involved in industrial tasks was developed by Jens Rasmussen of Denmark.[1] This pattern provides a useful framework for identifying the types of errors likely to occur in different operational situations, or within different facets of the same task that may place different information-processing demands on the individual. The classification system is known as the skill-based, rule-based, and knowledge-based approach. The three classifications refer to the degree of conscious control exercised by the individual over their activities [2].

In the knowledge-based mode, the task is carried out by the human in an almost totally conscious fashion. This would occur if a beginner (e.g., an operator in training) is performing a task, or if an experienced individual encounters a completely novel situation. In either of these circumstances, substantial mental exertion would have to be asserted to evaluate the condition, and their responses would likely be slow. In addition, after each action, the person would need to evaluate its effect prior to taking additional action, which would probably further slow their responses to the situation. Knowledge-based performance results in a nominal error rate of 1:2 [2].

In the skill-based mode, efficient performance of well-practiced, mainly physical actions of which practically no conscious reasoning occurs. Skill-based actions are normally commenced by an explicit occurrence, such as the requirement to operate a valve, that may arise from an alarm, a procedure, or an indication from another individual. The well-practiced task of opening the valve will then be executed largely without conscious thought. The skill-based performance mode results in a nominal error rate of 1:1,000 [2].

The last category involves the use of rules, which may have been learned as a result of interacting with the plant, through formal training, or by working with experienced process workers. The level of conscious control is midway between that of the knowledge- and skill-based modes. The rule-based performance mode results in a nominal error rate of 1:100 [2].[2]

Next, it is important to describe and distinguish between slips and mistakes. Slips are defined as errors in which the intention is correct, but a failure occurred in the actual carrying out of the activities required. For example, a worker may know that a receptacle needs to be filled but instead may fill a similar receptacle nearby. This slip may occur if the receptacles are poorly labeled, or if the worker is confused with regard to the location of the correct receptacle. Mistakes, by contrast, arise from an incorrect intention, which leads to an incorrect action sequence that may be quite consistent with the wrong intention. For example, a worker might wrongly assume that a reaction was endothermic and might apply heat, thereby causing overheating. Incorrect intentions may arise from lack of knowledge or an inappropriate diagnosis [2].

Slips can be described as due to misapplied competence because they are examples of errors in highly skilled, well-practiced activities that are characteristic of the skill-based mode. Mistakes, however, are largely confined to the rule- and knowledge-based performance modes [2].

In the skill-based mode, the individual can function very effectively by using pre-programmed sequences of behavior that do not require much conscious control. Checking on progress at specific points is only occasionally needed when operative in this mode. An undesirable consequence accompanying this efficiency is that strong habits can take over when attention to checks is diverted by distractions, or when unfamiliar activities are embedded in a familiar context [2].

With regard to mistakes, two separate mechanisms operate. In the rule-based mode, an error of intention can occur if an improper diagnostic rule is used. For example, a worker who has considerable experience in stagnant, shutdown-status power plant chemistry may have learned diagnostic rules that are inappropriate for operational, dynamic, and volatile power plant chemistry. If they attempt to apply these rules to evaluate the cause of a continuous process disturbance, a misdiagnosis could result, leading, in turn, to an inappropriate action. In other situations, diagnostic rules that have been successful in the past may be overused. Such sound rules are usually applied first even if they are not necessarily appropriate [2].

People often have a tendency to force a new situation into the mold of previous events. For example, in one incident, some modifications were made to a pump used to transfer a liquid. When movement of the liquid was complete, the worker pressed the stop button on the control panel and saw that the "pump running" light went out. The worker also closed a remotely operated valve in the pump delivery line. Several hours later, the high-temperature alarm on the pump sounded. Because the worker had stopped the pump and had seen the "pump running" light go out, the worker assumed that the alarm was faulty and ignored it. Shortly thereafter the pump exploded. The explanation for this unwanted sequence of events is that when the pump was modified, an error was introduced into the circuit. As a result, pressing the stop button did not stop the pump but merely switched off the running light. The pump continued running and overheated, and the material in it decomposed explosively. In this example, a major contributor to the accident was the worker's assumption that when the "pump running" light went out, the pump must have stopped. That assumption prevailed even through the sounding of a high-temperature alarm, which would usually be associated with an operating pump. The rule, "If the pump light is extinguished, then pump is stopped," was so strong that it overcame the evidence from the temperature alarm that the pump was still running [10].

In the case of knowledge-based mistakes, other factors are important. Most of these factors result from the considerable demands on information-processing capabilities that become necessary when a situation must be evaluated in unfamiliar conditions. Given these demands, it is not surprising that humans do not perform very well in high-stress, unfamiliar situations where they are required to think on their feet in the absence of rules, routines, and procedures to give them suitable direction. For example, operators may only use the finite information that is immediately available to evaluate the situation, rather than seeking more comprehensive data or assistance from others that are more knowledgeable. They may also become overconfident in the correctness of their knowledge. One typical behavior that occurs during knowledge-based problem solving is an insistence that one course of action is correct, leading an individual or the operating team to become tangled in one aspect of the problem and exclude all other aspects that should be considered. This behavior characterized the Three Mile Island nuclear accident in Pennsylvania. The opposite form of behavior can also be observed, in which the overloaded worker gives their attention superficially to one problem after another without solving any of them [2].

In the skill-based mode, recovery is usually prompt and effective because the individual will have familiarity with the expected outcome of their actions and will therefore get timely feedback with respect to any slips that may have prevented this outcome from being reached. This highlights the role of feedback as a significant aspect of error recovery. In the case of mistakes, the mistaken intention tends to resist contrary evidence. People tend to ignore feedback information that does not support their expectations of the situation [2].

ACTIVE ERRORS

Different types of errors can be categorized as active, latent, or foreseeable. Active errors have effects that are noticed immediately. They can occur across the spectrum of human behavior modes but are usually associated with individuals in frontline operations of a system, such as officers of ships, air traffic controllers, pilots, and control room operators in a nuclear facility. When examining active errors, we must take into account the complexity associated with human nature, which includes all the emotional, mental, social, biological, and physical characteristics that define people's limitations, abilities, and tendencies. An aspect of human nature relevant here is the innate tendency toward imprecision. Whereas machines tend to be precise, people are usually imprecise, especially when under stresses such as time pressure. Human fallibility can cause people to get into situations that are beyond their abilities. Logically, complex systems intensify a person's susceptibility to make mistakes [2].

Because active errors are common and often quite consequential, their most prevalent causes should be understood to aid in reducing them. For example, most individuals tend to overestimate their abilities to maintain control at their workstation.[3] In this instance, the maintenance of control means that the task occurs as it is supposed to with the person performing in the appropriate fashion. Such overestimation can occur for at least two reasons. First, consequential error is rare, and many times an error occurs with no adverse result. Thus, people conclude that errors will not be caught unless they are inconsequential. Second, people do not know or acknowledge their own capabilities. For example, most people can function on insufficient sleep or work during times of distraction. They can also perform work duties during poor environmental conditions such as extreme cold, heat, vibration, or noise. People can become accustomed to these conditions. However, if the limits of a person's capabilities are exceeded, the chance of making errors increases. The impact of physical or environmental limitations can be especially problematic when work is taking place within a complex system [2].[4]

Stress is a prominent contributor to active errors. Stress is not always a problem; sometimes, it is healthy and normal. Stress can focus attention and enhance an individual's performance. However, elevated stress can overpower an individual and thus become detrimental to performance. Stress can be understood as the body's physical and mental response to perceived threats within the environment. The important word in this case is "perceived" because the individual's perception is central to adaptation in order to cope with a threat. Stress tends to increase in conjunction with lack of familiarity with the situation. Extreme stress can lead to panic, which inhibits a person's ability to act or to sense, recall, or perceive essential elements of a situation. Fear and anxiety often follow when an individual believes that they cannot respond appropriately to a situation. This fear and anxiety are frequently accompanied by a lapse of memory and an inability to perform certain actions or to think critically [2].

Another important factor in mental errors is people's tendency to avoid mental strain. Most people engage only reluctantly in long periods of concentrated thinking. They also tend to avoid situations in which they must display heightened levels of attention for an extended period of time. Thought can be a slow and laborious process that requires significant effort. Therefore, people often seek familiar patterns and tend to apply solutions with which they are already familiar. This shortcutting is a type of mental bias designed to reduce the cognitive effort required in making decisions [2].[5]

One of these mental biases is assumptions. People frequently accept as true certain conditions that have not been verified. Another bias is habit, or an unconscious behavior pattern acquired through frequent repetition. Confirmation bias can also be problematic and is exemplified by a reluctance to abandon established solutions. Individuals tend not to change their way of thinking or behaving, even when there is conflicting information or when better solutions are available. Thus, people often defend their established position and ignore blatant evidence to the contrary. Next, frequency bias refers to a tendency to gamble that a familiar solution will work or to viewing information as more important when it has occurred more frequently. Finally, people often suffer from availability bias, the tendency to use solutions that immediately come to mind or to place greater importance on facts that are readily available [2].

Limited working memory can be a factor in active errors.[6] We rely on short-term memory to make decisions and solve problems. This short-term memory can be understood as a storeroom that demands attention and is temporary in nature. It is used to recall new information and is actively involved with recall, storage, and learning. When the limits of this memory are exceeded, errors can result [2].

LATENT ERRORS

As with active errors, latent errors may lead to adverse consequences, but these may lay dormant within a complex system for a significant period of time before they manifest. They often become evident only when combined with other factors to result in a breach of a system's defenses. Latent errors are frequently committed by individuals whose activities are removed in space and time from the direct human system interface. For example, they may be committed by maintenance personnel, managers, construction workers, high-level decision-makers, or system designers well before their manifestation [2].

An analysis of significant nuclear accidents such as Chernobyl and Three Mile Island found that latent errors frequently pose the most important threat when people interact with a complex system.[7] Traditionally, accident investigations and reliability analyses have concentrated on direct equipment failures and operator errors. Although operators do make mistakes, such as those presented above as examples of active errors, many of these mistakes have an underlying cause connected to a latent error, such as when an operator of a complex system inherits the mistakes made by the designers or installers [2].

For these reasons, the study of latent failures may be more beneficial than a focus on operator mistakes. Unfortunately, most research on human factors has concentrated on improving the human-machine interface, an emphasis that entails a focus on active errors, even though latent errors inherent within the system can be associated with a broader range of possible problems. In other words, active errors may be only the outcome of latent problems that have long been embedded within the system [2].

Latent human error comes into play when individuals' propensity for error is enhanced by the environment in which they work and the systems with which they interact. Two adverse effects can result from latent conditions: their ability to provoke errors, and their impact on the long-term health and welfare of the system that created them. These conditions do not necessarily contribute immediately to the possibility for error; rather, they can rest hidden within a system until the requisite elements align and cause the latent error to become activated [2].

Rapid technological advances across all industries have generated an additional focus on latent error. Many of today's complex systems have operators who are remote and removed from the processes that they control. As the systems have become more complex, they can stand between people and the physical tasks involved. When nuclear technology was initially introduced, operators still engaged in direct manipulation and sensing of the systems that they were operating. In other words, they could still touch and see the system that they controlled. As technology has continued to advance, the remote manipulation of devices and sensing has further removed humans from the processes under their charge [2].

The most significant changes in how humans interact with complex systems have resulted from the decreased cost of powerful computing. Many system operators are now separated from their process by more than one component of a control system. At a lower level, the task interactive system controls the detailed parts of an operation. There is an intervention between the specialized system and the operators due to the need for a human system interface. The control system presents pieces of information to the operator, but the interface allows only a prescribed degree of interaction between the person and the remote process. This creates a situation of supervisory control. The person adjusts, monitors, and initiates processes and systems that are also automatically controlled [2]. Nevertheless, the stimuli contained within the operational environment are always impacting these remote operators and can still contribute to errors. Moreover, their remoteness can increase the possible impact of latent errors introduced by people who designed or installed the system. In fact, those nuclear manufacturers, installation teams, and facility personnel who are considered the best in their field may be prone to making the worst mistakes [10]. Although the focus in accident investigation may be placed upon the operators, another key source of errors arises when the systems themselves are not scrutinized for their own propensity to cause errors [2].[8]

The increasing complexity and automation of systems at nuclear facilities are both making latent errors more difficult to detect and giving them greater capability to lead to a serious incident. If a latent error occurs in the development of a monitoring system, a remote operator may not become aware of any problems until it is too late to reverse the process. Alternatively, a latent error could be embedded in the design of a semi-automated control process. In this case, the operator might detect a problem with a plant parameter and might initiate the necessary actions to correct the problem. However, if a latent error is present, the control processes may not respond appropriately to the operator's actions. As a result, alternative means costing more time and money may be required, or the severity of the problem could escalate [2].

These problems are related to the design of a monitoring and a control system, respectively. Unlike an active human error, such design problems could span numerous pieces of equipment and could potentially go undetected for years, even decades. Not until a problem in plant operation arises would such a latent error become detectable. For this reason, it is crucial for safety monitors and researchers to shift away from the traditional approach of concentrating on operator errors. Although active errors are important, latent errors can be even more catastrophic.

FORESEEABLE ERRORS

Human beings are notoriously error-prone, and we now know a great deal about the specific types of errors that people are likely to make. These mistakes are frequently described as foreseeable errors. Because they are expected, they can be accounted for in a system. Complex modern systems can be designed to accommodate this category of human errors and ensure that they remain benign. This approach is called "designing for error" [11]. As technology advances and complexity expands, the number of foreseeable errors tends to rise, requiring additional protective steps.

A number of factors must be considered when designing a complex system to accommodate foreseeable human error [12]. One way to accomplish this is to do a human reliability assessment (HRA), which encompasses techniques for statistically determining the probability that a complex system will fail given certain circumstances. The techniques take into account the chance that various components of the system, including the operators as well as mechanical components, will fail in specific ways [13]. In this way, the human operator is considered a part of the system that is fallible like any other.

Human performance varies across individuals and situations [14]. Tasks require different levels of manual skills, training, and attention. As I noted above, the types of work in which people engage can be broken down into three general categories: knowledge-based, rule-based, and skill-based. Using these categories as constraints assists in the analysis of human behavior and error [15].

The probability of human error is proportional to the necessary level of knowledge needed to perform the task [14]. Tasks that require a complicated series of actions must frequently be done in a specific order. For this to happen, an individual must have the knowledge ahead of time or must be provided with it at a key moment. Information that must be maintained in the operator's memory is sometimes known as "knowledge in the head." Processes that involve high levels of knowledge in the head also tend to result in a significant number of errors [13].

Along with knowledge in the head, there is "knowledge in the world," which refers to knowledge held within the components of the task [16]. In that situation, the elements included in the task contain information necessary for the proper performance of the task. For example, in the case of an assembly that involves several washers and nuts, there may be only one sequence that results in the proper disassembly of these components. However, the disassembly sequence is contained within the elements of the task. These types of situations result in fewer errors than those requiring knowledge in the head, presumably because they do not depend on the reliability of human memory [14].

Another approach to designing a system that can accommodate foreseeable human error is the critical incident technique, which assumes that problems do not occur spontaneously [17]. Each failure is associated with critical incidents that allow the failure to occur. A critical incident is a situation in which the errors nearly cause a failure, or a failure is already occurring but something prevents a disaster from ensuing. The critical incident technique is an anonymous reporting tool that usually relies on a survey to elicit information from operators. This approach works most efficiently when it is part of a continuing program, not a single solicitation. The people supplying information for a critical incident program are generally encouraged to provide sufficient identification information so that they can be contacted by the investigators if necessary. However, this request for personal disclosure can discourage response. In all cases, for proper implementation of this specific technique, the individuals providing information must have their identity removed from any reports so that other individuals at their place of employment cannot identify them [18].

An effective way to design a system that accounts for foreseeable human errors is a fault tree analysis, which is diagrammatic and requires that the system be logically broken down into its functional components [19]. The relationships among these components are then identified in a diagram that depicts the elements and their relationships to each other. Most fault trees have two possible outcomes: a specific type of failure or a successful operation. These possibilities are represented at the top of the tree. The elements that result in each outcome are listed below. The branches of the tree illustrate the logical relationships between the functional elements. Some fault tree analyses will include a probabilistic risk assessment that calculates the statistical probabilities associated with each branch of the tree. This approach to accounting for foreseeable human errors has the advantage of being a top-down analysis that begins with the postulate of an outcome, such as, in the case of a nuclear power plant, a reactor core breach [16].

UNEXPECTED EVENTS

In contrast to the others, unexpected events are a type of abnormal behavior of systems themselves [20]. They are often caused by human errors and result in the loss of productivity. Many of the more popular practices and methodologies applied in systems engineering do not reduce the chance of unexpected events. However, engineers can reduce these operational risks if they take into account the limitations of human operators when interacting in a system [21]. As with foreseeable errors, as technology advances and complexity grows, the possibility of unexpected events will rise and require added analysis and remediation.

It is clear that human limitations provide numerous chances for unexpected events to occur. Usability engineering can frequently solve some of the problems, but this approach requires significant knowledge of system behavior to integrate safeguards [18]. The methods used to mitigate risks of human error must be integrated within the system through engineering practices. This involves the addition of guidelines and methods to traditional systems engineering to protect the system from unexpected events. These guidelines are based on a study of the known failures, interviews with systems engineers, and background research.

Once again, we can use classification to help us understand unexpected events, such as operator confusion, intra-system inconsistency, and inter-system mismatches. Operator confusion occurs when there is a mismatch between the operator's perception of the state of the system and its actual state; intra-system inconsistencies include problems that occur between units within a system; and inter-system mismatches are problems resulting from faulty communication between different systems [22].

During standard operations, the system units will comply with common scenarios [23]. This is often referred to as a "normal system state." When operating, the system units are receiving event information. An event is considered normal if a unit is designed to respond properly to the specific event. If not, the event is described as a "slip." Many slips are caused by hardware faults or an unexpected action by the user. The occurrence of a slip changes the system to an exceptional state. A resilient system will contain protocols to return the system to its normal state. In a non-resilient system, the next event may result in some type of mishap or unexpected outcome, creating an unexpected event [18].

As most systems are designed for operation according to certain scenarios, the responses to an event by any of a system's units are designed according to a specific operating scenario based on various assumptions. During normal operations, a system assumes a certain scenario, providing a system context. When a system unit receives input indicating an exceptional event, the operating scenario may be different. For example, a unit failure may result in an operating scenario that changes to call for the unit's replacement. If the system's units all operate according to the changed scenario, then the system is described as being context compliant. If this does not occur and some of the system's units are not complying with the new context, then the system enters a state of being context-inconsistent [21].

Events can be understood as expected if they are in compliance with the operating scenario, which defines the context of operation. For all the system units to work properly, they need to behave in compliance with the procedures that have been defined in the context. This includes the operator. Unexpected events result from the need for human control in exceptional situations such as an abnormal event, alert, or emergency. During design, the primary scenarios are anticipated, but it is impossible to predict all scenarios. To handle exceptional situations properly and safely, the system relies on a human operator. In such instances, the operator will have unusual control over the system. It is expected that human operators will use proper judgment when applying their exceptional control [23].

Exceptional events have a wide range of sources, including designs that do not fit operational needs, interruptions in the normal operating procedure, unintentional actions that are not compliant with their context, failure to comply with a changing context, and false perceptions [22].

TAXONOMIES

While such classifications are helpful, a taxonomy classification method to classify and code errors is essential to analyze data and spot error patterns within a system's design. However, taxonomies are industry-specific, and the classifications that work for one industry may not work for another. Also, when taxonomies are applied in a nuclear setting, no single taxonomy can cover all possible scenarios, and taxonomies may change with new data. Systems must be designed with this type of rapid evolution in mind.

A traditional system design is based on the assumption that certain required specifications will be met. These specifications are based, in turn, on a requirement analysis performed in accordance with defined scenarios. In practice, however, specification documents are often not well associated with conceivable scenarios. Furthermore, most documents on procedure specifications do not describe the relationships within the system's states. They may also not describe the appropriate desired responses to all possible events. As a consequence, the system design derived from this input does not include the necessary means for matching the system's activity with the operating scenarios [20].

Moreover, it is common for a system's behavior to be properly understood only after an incident occurs. This means that an event that resulted in a mishap was not anticipated during the design. Frequently, designers do not or cannot consider all possible states of the system during the design stage. The more events that can be anticipated, the more can be handled by the system to prevent negative outcomes. Proper protection against unexpected events requires formalizing them within operational scenarios. There must also be a close relationship between the state of the system and the scenario [21].

NOTES

1 Jens Rasmussen (May 11, 1926–February 5, 2018; born in Ribe, Denmark) was a system safety and human factors professor at Risø in Denmark. He was highly influential within the field of safety science, human error, and accident research, contributing with the skill rule knowledge framework, Risk Management Framework, AcciMap Approach, and others.

2 For more, see James Reason, *Human Error* (Cambridge, MA: Cambridge University Press, 1990).

3 For more, see P. C. Cacciabue, and M. Cassani, "Modelling Motivations, Tasks and Human Errors in a Risk-Based Perspective," *Cognition, Technology & Work* 14, no. 3 (2011): 229–41, http://doi.org/10.1007/s10111-011-0205-4.

4 For more, see John Stoop, and Sidney Dekker, "Are Safety Investigations Pro-Active?," *Safety Science* 50, no. 6 (2012): 1422–30, http://doi.org/10.1016/j.ssci.2011.03.004.

5 For more, see Paul S. Goodman, Rangaraj Ramanujam, John S. Carroll, Amy C. Edmondson, David A. Hofmann, and Kathleen M. Sutcliffe, "Organizational Errors: Directions for Future Research," *Research in Organizational Behavior* 31 (2011): 151–76, http://doi.org/10.1016/j.riob.2011.09.003.

6 For more, see E. Rebhan, "Challenges for Future Energy Usage," *European Physical Journal Special Topics* 176, no. 1 (2009): 53–80, http://doi.org/10.1140/epjst/e2009-01148-9.

7 For more, see Michael E. Stern, and Margaret M. Stern, "Electric Power in a Carbon Constrained World: Does Nuclear Power Have a Future?," *Utah Environmental Law Review* 32 (2012): 431–89.
8 For more, see James Reason, "Human Error: Models and Management," *BMJ* 320, no. 7237 (2000): 768–70. http://doi.org/10.1136/bmj.320.7237.768.

REFERENCES

1. Republished with permission of American Society of Mechanical Engineers, from Corrado, Jonathan. 2022. "Proactive Human Error Reduction Using the Systems Engineering Process." *ASME Journal of Nuclear Engineering and Radiation Science* 8(2). doi:10.1115/1.4051362; permission conveyed through Copyright Clearance Center, Inc.
2. Republished with permission of American Society of Mechanical Engineers, from Corrado, Jonathan. 2021. "The Intersection of Advancing Technology and Human Performance." *ASME Journal of Nuclear Engineering and Radiation Science* 7(1). doi:10.1115/1.4047717; permission conveyed through Copyright Clearance Center, Inc.
3. Karwowski, Waldemar. 2006. *International Encyclopedia of Ergonomics and Human Factors*. 2nd ed. Boca Raton, FL: CRC Press/Taylor & Francis Group.
4. Salvendy, Gavriel. 2012. *Handbook of Human Factors and Ergonomics*. 4th ed. Hoboken, NJ: John Wiley & Sons.
5. O'Connor, Paul E., and Joseph V. Cohn. 2010. *Human Performance Enhancement in High-Risk Environments: Insights, Developments, and Future Directions from Military Research*. Santa Barbara, CA: Praeger.
6. Matthews, Gerald, Roy D. Davies, Stephen J. Westerman, and Rob B. Stammers. 2000. *Human Performance Cognition, Stress and Individual Differences*. Hove: Psychology Press.
7. Reiman, Michael P., and Robert C. Manske. 2009. *Functional Testing in Human Performance*. Champaign, IL: Human Kinetics.
8. Duffy, Vincent G., ed. 2009. *Handbook for Digital Human Modeling: Research for Applied Ergonomics and Human Factors Engineering*. Boca Raton, FL: CRC Press/Taylor & Francis Group.
9. Foyle, David C., and Becky L. Hooey, eds. 2008. *Human Performance Modeling in Aviation*. Boca Raton, FL: CRC Press/Taylor & Francis Group.
10. Reason, J. 1990. *Human Error*. Cambridge: Cambridge University Press.
11. Rebhan, E. 2009. "Challenges for Future Energy Usage." *European Physical Journal Special Topics* 176(1): 53–80. doi:10.1140/epjst/e2009-01148-9.
12. Cacciabue, P. C., and M. Cassani. 2011. "Modelling Motivations, Tasks and Human Errors in a Risk-Based Perspective." *Cognition, Technology & Work* 14(3): 229–41. doi:10.1007/s10111-011-0205-4.
13. Stoop, John, and Sidney Dekker. 2012. "Are Safety Investigations Pro-Active?" *Safety Science* 50(6): 1422–30. doi:10.1016/j.ssci.2011.03.004.

14. Goodman, Paul S., Rangaraj Ramanujam, John S. Carroll, Amy C. Edmondson, David A. Hofmann, and Kathleen M. Sutcliffe. 2011. "Organizational Errors: Directions for Future Research." *Research in Organizational Behavior* 31: 151–76. doi:10.1016/j.riob.2011.09.003.
15. Rasmussen, Jens. 1983. "Skills, Rules, and Knowledge; Signals, Signs, and Symbols, and Other Distinctions in Human Performance Models." *IEEE Transactions on Systems, Man, and Cybernetics* 13(3): 257–66. doi:10.1109/TSMC.1983.6313160.
16. Ramanujam, Rangaraj, and Paul S. Goodman. 2011. "The Link between Organizational Errors and Adverse Consequences: The Role of Error-Correcting and Error-Amplifying Feedback Processes." Chap. 10 in *Errors in Organizations*, edited by David A. Hofmann Michael Frese, 245–72. New York: Routledge/Taylor & Francis.
17. Flanagan, John C. "The Critical Incident Technique." *Psychological Bulletin* 51, no. 4 (1954): 327–58.
18. Mengolini, Anna, and Luigi Debarberis. 2012. "Lessons Learnt from a Crisis Event: How to Foster a Sound Safety Culture." *Safety Science* 50(6): 1415–21. doi:10.1016/j.ssci.2010.02.022.
19. Stern, Michael E., and Margaret M. Stern. 2012. "Electric Power in a Carbon Constrained World: Does Nuclear Power Have a Future?" *Utah Environmental Law Review* 32: 431–89.
20. Dauer, L. T., P. Zanzonico, R. M. Tuttle, D. M. Quinn, and H. W. Strauss. 2011. "The Japanese Tsunami and Resulting Nuclear Emergency at the Fukushima Daiichi Power Facility: Technical, Radiologic, and Response Perspectives." *Journal of Nuclear Medicine* 52(9): 1423–32. doi:10.2967/jnumed.111.091413.
21. Lelieveld, J., D. Kunkel, and M. G. Lawrence. 2012. "Global Risk of Radioactive Fallout after Major Nuclear Reactor Accidents." *Atmospheric Chemistry and Physics* 12(9): 4245–58. doi:10.5194/acpd-12-19303-2012.
22. Wehrden, Henrik von, Joern Fischer, Patric Brandt, Viktoria Wagner, Klaus Kümmerer, Tobias Kuemmerle, Anne Nagel, Oliver Olsson, and Patrick Hostert. 2012. "Consequences of Nuclear Accidents for Biodiversity and Ecosystem Services." *Conservation Letters* 5(2): 81–89. doi:10.1111/j.1755-263x.2011.00217.x.
23. Mohan, M. P. Ram. 2011. "Nuclear Energy: Nuclear Safety." *Yearbook of International Environmental Law* 22(1): 219–24. doi:10.1093/yiel/yvs081.

4 An Observational Study on the Interface between Human Error and Technology Advancement

RESEARCH DESIGN

With a better understanding of what I studied for this book, I can turn to explaining how I studied it, which was observationally. As such, I have drawn inferences from the effect of certain variables (technological changes) on a situation (nuclear reactors with accidents). While, as usual, I have used a treatment group and a control group, there was no way for me to assign the members of this group randomly. I instead grouped nuclear facilities into those that have had accidents as a result of technological changes that affected the way the operator interacted with the system (treatment group) and those that have had accidents without recent technological changes (control group) [1].

It is not feasible for any researcher to conduct an experiment that would involve requiring nuclear facilities to use certain technologies. Therefore, the effect of these changing technologies on the facility must be observed after the fact. The observational study is an excellent method of collecting real-world information. In this case, it is the most suitable way to assess evidence of the

DOI: 10.1201/9781003346265-4

risks related to the introduction of new technologies at nuclear facilities. The observational study can be used to formulate hypotheses that can be tested in other experiments. It is also a good way to generate data that can be used to formulate effective policies and safeguards for the nuclear industry [2].

Because I had to use a non-randomized approach regarding the application of the treatment of new technology, I had to account for potential experimental bias. For example, it could be that the facilities that underwent technology changes were older on average, with the result that if these facilities had a higher prevalence of incidents, facility age could be an alternative explanation rather than the introduction of the new technology. I compensated for that by examining the data carefully, anticipating potential problems, and attempting to gauge their impact.

THEORETICAL FRAMEWORK

Considering the challenges I faced, I made use of several theoretical frameworks from multiple disciplines, including human factors and ergonomics (HF&E), which itself incorporates insights from numerous disciplines, including anthropometry, operations research, statistics, graphic design, industrial design, mechanobiology, biomechanics, engineering, and psychology [3]. The field of HF&E emerged during World War II, at a time of considerable development in the complexity of weaponry and other machines, which placed great demands on operator cognition. During the Information Age, HF&E has been extended to human-computer interactions [4].

HF&E is now used in a wide variety of fields including virtual environments, training, transportation, product design, information technology, healthcare, geriatrics, aerospace, and nuclear facilities [4]. For the last, HF&E seeks to reduce the strain placed on operators to decrease the amount of cognitive and physical energy they must put forth in order to operate nuclear facilities [3].

I also employed the methodology of sociotechnical systems, which is a type of organizational development that recognizes that the interactions occurring between technology and people at the workplace can be optimized through often complex organizational work [5]. The term can also be used to describe the intricate infrastructures in society and how they interact with human behavior. Both definitions are applicable to nuclear facilities, as both elaborate infrastructures and complex interactions between people and advanced technological systems that are involved in the proper operation of a nuclear facility [6].

Sociotechnical systems represent a type of theory regarding society, people, and the technical processes and organizational structures involved in complex systems [7]. The term "technical" does not always refer to material technologies; it can also describe knowledge and procedures. "Sociotechnical" emphasizes that an organization often involves goals and tasks that are both technical and social in nature. The sociotechnical analysis of a system strives to achieve a joint optimization of both the social aspects and technical performance of the system. This is especially helpful with nuclear facilities, which function in a social environment. For example, regulatory requirements that govern nuclear facilities and affect the systems with which the operators must interact are highly technical yet have a social aspect because they must satisfy the demands of multiple stakeholders [6].

Probabilistic risk assessment (PRA) is a relatively comprehensive, systematic methodology used to evaluate the risks associated with complex technological entities such as nuclear facilities [8]. In these cases, the risk is understood as a potential detrimental outcome of an action or activity. Two quantities are associated with risk: the likelihood of an occurrence and the magnitude of the adverse effects. These are often described as the probability of the problem and its severity. Consequences are generally expressed numerically by multiplying the probability times the total risk to calculate the expected loss [9].

PRA is often used to assess nuclear facilities through three steps. First, one must identify the problems that can occur at the facility and whether they could lead to an adverse consequence. Second, one must assess the severity of the potential problem, in terms of their possible consequences for the facility. The third item to be examined is the likelihood that these consequences will occur, typically described as a frequency or probability. Common methods of conducting a PRA are fault tree analysis and event tree analysis. PRAs generally fall under the classification of safety engineering [10].

User-centered design, a subdivision of user-interface design pays attention to each stage in the design process to accommodate the limitations and needs of end users. This is a multistage process of problem solving that requires designers to foresee and analyze how human users of the technology are likely to interact within a system. The process involves testing in a real-world context the validity of assumptions with regard to users' behavior. This type of testing is necessary because it can be nearly impossible for product designers to understand the experience of a first-time user of the technology. One must also consider the learning curve of people interacting within the system. All these factors are crucial in the efficient, proper operation of a nuclear facility. User-centered design is critical for the proper organization of workstations, other types of interaction between human users, and complex systems involved with a nuclear facility [10].

Human reliability assessment (HRA) is concerned with evaluating the probability that human error will occur when a specific task is completed. This type of analysis informs actions aimed at reducing the likelihood of such errors and improving the overall safety system. This type of analysis is only one of many performed at nuclear facilities to ensure that incidents are minimal [11].

HRA has three main goals: error reduction, quantification, and identification [11]. Several techniques are used for these purposes, including the technique for human error rate prediction (THERP). The techniques can be categorized as either first- or second-generation. First-generation techniques are based upon a dichotomy of either fitting or not fitting the error situation within a given context; second-generation techniques are based on theory and serve to quantify the errors. HRA techniques are used in a number of disciplines, including business, transportation, general engineering, healthcare, and nuclear operation.

The US Nuclear Regulatory Commission (NRC) contracted with the author of THERP to design a more consistent tool for determining human error rates at nuclear facilities. This resulted in the development of the accident sequence evaluation program human reliability analysis procedure (ASEP), which tends to be more conservative than THERP. One significant advantage of the ASEP is that it does not require the user to be an expert in human factors engineering. In addition, the training needed to use the ASEP is relatively minimal. After it became computerized, ASEP began to be referred to as simplified human error analysis code (SHEAN) [11].

Latent human error, as discussed in Chapter 3, refers to human error due to systems or routines formed in such a way that humans are disposed to make these errors [3]. This term is widely used throughout the aviation industry, but it is also prevalent in the safety literature for nuclear facilities. When this method is used at a nuclear facility, operator error data are gathered, grouped, collated, and analyzed to provide information for determining whether a disproportionate number of errors is present in a given system piece. If errors at a particular point in the system are too numerous, the system or routine can be analyzed, the potential for the problems identified, and changes made. This decreases the likelihood of future errors that could result in nuclear incidents [12].

RESEARCH METHOD

I used the International Nuclear and Radiological Event Scale (INES), as discussed in Chapter 2, to determine whether an event at a nuclear facility reached a sufficient level of severity to be considered. Though they are useful, ratings on the INES are also subjective. Because nuclear facilities are manmade and vary greatly in design, incidents occurring at these plants are subject to certain interpretations when one attempts to estimate the magnitude of the problem. I considered only incidents rated at Level 3 or higher as sufficiently severe to be included. In the United States, NRC, which carries responsibility for atomic safety in this country but cooperates with the International Atomic Energy Agency (IAEA), uses a nuclear event scale similar to the INES and transfers its ratings to the INES for reporting purposes. I used ratings from the INES after the numbers were converted [1].

The information I used was available on regulatory agency databases and in books detailing the investigation results of the incidents. Available information on incidents at Level 3 or higher and plant data from the time period before the incident occurred were gathered to determine the level of technological advancement impacting human operators at the facility.

The IAEA and NRC both report on events that they have investigated. However, possibly due to the diverse nature of their cultures, policies, and relevant laws, the IAEA data are not as complete or detailed as those provided by the NRC. As I explained in Chapter 2, NRC information is converted to INES levels. I used the event notification reports of the NRC and its predecessor US regulatory agencies. This approach provided an opportunity to use a standardized system that assesses the severity of each nuclear event while taking advantage of the more detailed reports released by the NRC [2].

The time period examined was from 1955 through 2021. I used only those events reaching Level 3 and above for two reasons. First and most important, any incident below Level 3 is relatively minor and unlikely to cause death, injury, or a problematic release of radioactive material. Second, I wanted to work with a reasonable number of incidents. As the incident level decreases, the occurrences increase exponentially, and using incidents below Level 3 involves processing an overwhelming amount of data due to the excellent daily reporting criteria applied by the NRC. In fact, most days contain multiple reports of events below the Level 3 threshold [2].

DATA COLLECTION, PROCESSING, AND ANALYSIS

Accordingly, I started collecting data by looking at the event notification reports provided by the NRC. I identified all events at Level 3 or higher from 1999 to 2021. Information on serious events prior to 1999 was obtained from books addressing the topic, on the basis of which I determined whether the incident would have been classified as Level 3 or higher on the INES [2].

After identifying all US incidents meeting this criterion for the sixty-six-year time frame, I sought information online regarding technological changes at the plant shortly before the event. Examples of technological changes include system and/or component upgrades, user interface alterations, and control system changes. The events were then categorized as to whether they followed the implementation of technological changes that may have affected how personnel interacted with the facility's systems. Changes that did not affect interaction between people working at the facility and its systems were discounted. For example, changes in the speed with which information is passed between computers would not normally be likely to affect the performance of human operators. However, if such a change modified the way in which the operators interacted within the system, then this would be considered a relevant technological change [2].

I used the measure of the financial cost of the incident in US dollars for determining the deleterious effects of the nuclear incidents. While there are certainly other costs such as psychological impacts associated with these incidents, they are difficult to quantify and, therefore were not used in this study. The reports on US incidents contain cost estimates; accounting for inflation is necessary for comparison purposes, so all amounts are adjusted for inflation to 2005 dollars.

METHODOLOGICAL ASSUMPTIONS

All this assumes that the reporting of events by the regulatory agencies is accurate, both in terms of financial costs and providing enough information to determine reliably whether the incidents followed technological changes. Along with other efforts to follow ethical considerations and get the most out of the data,[1] I attempted to reduce bias in this regard by recruiting colleagues to assess available information on the incidents and assess whether, in their judgment, any relevant technological changes had occurred. I provided my colleagues with all the data that I had located and invited them to do further research if necessary to either corroborate or refute my assessment. There were no inconsistencies between my colleagues' determinations and my own, confirming that I was not inadvertently influencing the outcome of the study by placing incidents in the wrong group.

NOTE

1 This study involves data previously collected by nuclear regulatory agencies. Since the information is publicly available, there is no issue of voluntary participation. Since no individual subjects are involved with the research, informed consent is not an issue. The anonymity and confidentiality of individuals who may be involved in the reports have already been protected because the reports do not mention any individuals specifically. This protection also minimizes the potential for harm to specific subjects. The remaining categories of ethical issues to address are the communication of the results and any other ethical issues specific to this project. Appropriate communication of results is an important consideration. Every attempt has been made to account for all relevant information available to the study. This means that all information found was reported, even if it did not support the hypothesis. Furthermore, the data were reported accurately without exaggeration. I should also state that I have no ties to regulatory agencies or to any of the nuclear facilities examined.

REFERENCES

1. Republished with permission of American Society of Mechanical Engineers, from Corrado, Jonathan, and Ronald Sega. 2020. "Impact of Advancing Technology on Nuclear Facility Operation." *ASCE-ASME Journal of Risk and Uncertainty in Engineering Systems, Part B: Mechanical Engineering* 6(1). doi:10.1115/1.4044784; permission conveyed through Copyright Clearance Center, Inc.
2. Creswell, John W. 2013. *Research Design*. 4th ed. Thousand Oaks, CA: Sage.
3. Karwowski, Waldemar. 2006. *International Encyclopedia of Ergonomics and Human Factors*. 2nd ed. Boca Raton, FL: CRC Press/Taylor & Francis Group.
4. Proctor, Robert W., and Trisha Van Zandt. 2008. *Human Factors in Simple and Complex Systems*. 2nd ed. Boca Raton, FL: CRC Press/Taylor & Francis Group.
5. Vermaas, Pieter, Peter Kroes, Ibo van de Poel, Maarten Franssen, and Wybo Houkes. 2011. "A Philosophy of Technology: From Technical Artefacts to Sociotechnical Systems." *Synthesis Lectures on Engineers, Technology and Society* 6(1): 1–134. doi:10.2200/s00321ed1v01y201012ets014.
6. Neyer, Anne-Katrin, Angelika C. Bullinger, and Kathrin M. Moeslein. 2009. "Integrating Inside and Outside Innovators: A Sociotechnical Systems Perspective." *R&D Management* 39(4): 410–19. doi:10.1111/j.1467-9310.2009.00566.x.
7. Gorman, J. C., N. J. Cooke, and E. Salas. 2010. "Preface to the Special Issue on Collaboration, Coordination, and Adaptation in Complex Sociotechnical Settings." *Human Factors* 52(2): 143–46. doi:10.1177/0018720810372386.

8. Mohaghegh, Zahra, Reza Kazemi, and Ali Mosleh. 2009. "Incorporating Organizational Factors into Probabilistic Risk Assessment (PRA) of Complex Socio-Technical Systems: A Hybrid Technique Formalization." *Reliability Engineering & System Safety* 94(5): 1000–18. doi:10.1016/j.ress.2008.11.006.
9. Stamatelatos, Michael, and Homayoon Dezfuli. 2011. *Probabilistic Risk Assessment Procedures Guide for NASA Managers and Practitioners* (NASA/SP-2011-3421). 2nd ed. Washington, DC: National Aeronautics and Space Administration, Office of Safety and Mission Assurance.
10. U.S. Nuclear Regulatory Commission. 2011. *Regulatory Guide 1.174: An Approach for Using Probabilistic Risk Assessment in Risk-Informed Decisions on Plant-Specific Changes to the Licensing Basis.* Washington, DC: Division of Risk Analysis and Applications, Office of Nuclear Regulatory Research, U.S. Nuclear Regulatory Commission.
11. Tonţ, Gabriela, Luige Vlădăreanu, Radu Adrian Munteanu, and Dan George Tonţ. 2009. "Some Aspects Regarding Human Error Assessment on Resilient Socio-Technical Systems." In *WSEAS International Conference on Mathematical and Computational Methods in Science and Engineering, Baltimore, MD, November 7–9, 2009*, 139–44. Stevens Point, WI: WSEAS.
12. Salvendy, Gavriel. 2012. *Handbook of Human Factors and Ergonomics.* 4th ed. Hoboken, NJ: John Wiley & Sons.

The Data Analysis and Result Interpretation Correlating Human Error and Technology Advancement in Nuclear Operation

5

DATA ANALYSIS

I used the *t*-test to determine whether a significant difference between the two groups (incidents determined to be caused by human error as a result of technological advances and those determined not to be caused by human error as a result of technological advances) existed. An independent samples *t*-test determines whether two independent groups are different from each other, and the two groups are not related, so an independent samples *t*-test can be used in this case. Before conducting the *t*-test, since it is a parametric test and requires normally distributed data, the outcome variable (costs of incidents) was checked

DOI: 10.1201/9781003346265-5

for normality and homoscedasticity of variance by examining the distribution of the data using histograms and Q-Q plots [1].

From 1955 to 2021, fifty-three incidents were determined to meet the INES Level 3 and above criteria and had an average incident cost of $190,518,900 (SD = $461,511,400). Among the fifty-three incidents, twenty-two were determined to be due to human error as a result of technological advances and thirty-one were not. Of this latter group of thirty-one that did not include human error, the cost of the incidents ranged from $1,000,000 to $695,000,000, with an average of $54,274,200 (SD = $142,559,800). For the twenty-two incidents due to human error as a result of technological advances, the cost ranged from $2,000,000 to $2,483,000,000 with an average of $382,500,000 (SD = $657,543,610) [1].

For the first assumption of the *t*-test, normality of data, I used the Kolmogorov-Smirnov test to compare the scores in the sample to a normally distributed set of scores. If this test result is significant, then the distribution of scores of this sample is significantly different from a normal distribution of scores. The Kolmogorov-Smirnov test was significant ($p < 0.001$). Thus, the first assumption of the *t*-test was violated.

A Q-Q plot of the scores also indicated that the data were not normally distributed (see Figure 5.1). A histogram (see Figure 5.2) reveals that

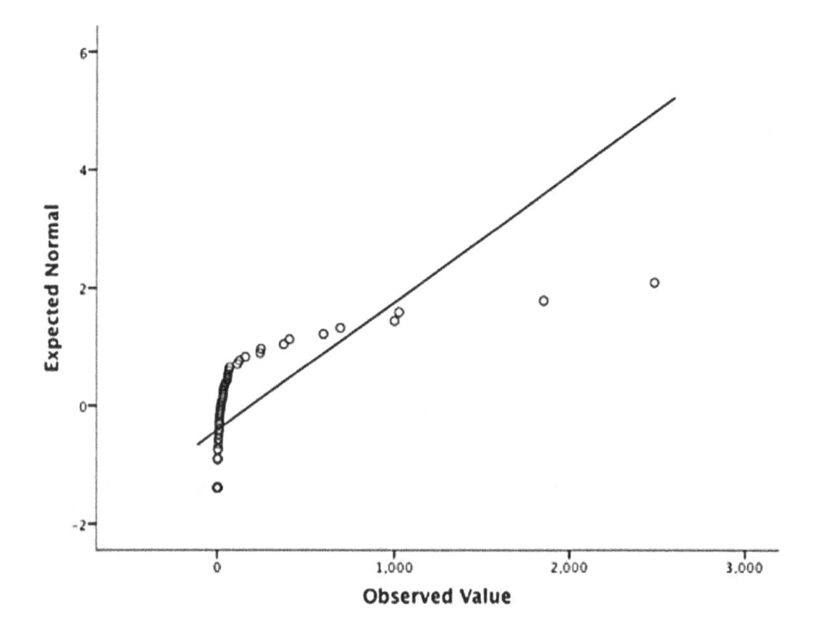

FIGURE 5.1 Q-Q Plot of Cost [Reproduced with Permission of American Society of Mechanical Engineers, from "Impact of Advancing Technology on Nuclear Facility Operation," *ASCE-ASME Journal of Risk and Uncertainty in Engineering Systems, Part B: Mechanical Engineering*, Corrado, Jonathan and Ronald Sega, 6(1), 2020; Permission Conveyed through Copyright Clearance Center, Inc.].

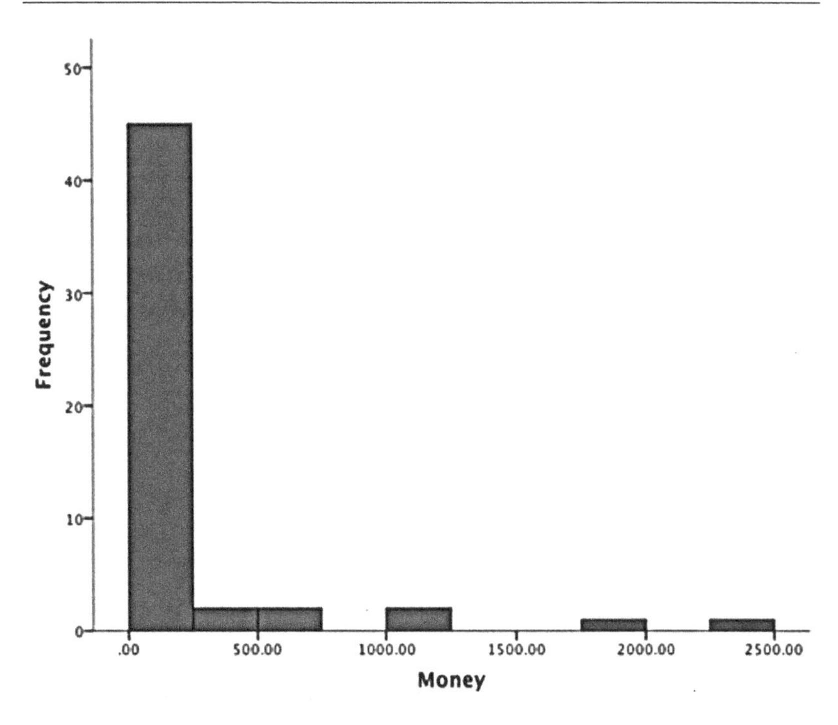

FIGURE 5.2 Histogram of Cost [Reproduced with Permission of American Society of Mechanical Engineers, from "Impact of Advancing Technology on Nuclear Facility Operation," *ASCE-ASME Journal of Risk and Uncertainty in Engineering Systems, Part B: Mechanical Engineering*, Corrado, Jonathan and Ronald Sega, 6(1), 2020; Permission Conveyed through Copyright Clearance Center, Inc.].

the data were positively skewed. Therefore, the data were log-transformed. Log-transforming data can make skewed data more normally distributed (see Figures 5.3 and 5.4) [1].

A Q-Q plot is a form of scatterplot created by plotting two sets of quantiles against one another to assess normality of the data. Quantiles are not the same as the actual observations and different ranges of values can be presented relative to histograms. When the data contain extreme values, as in this study, these may not be rendered in the graph as seen in the histogram (as displayed in Figures 5.3 and 5.4). Thus, it is not expected that the exact ranges of values should be consistent between histograms and Q-Q plots since they plot different factors: quantiles for Q-Q plots and actual values for histograms [1].

Using the log-transformed data, I conducted an independent samples *t*-test and checked the normality of the data using the Kolmogorov-Smirnov test. This time, the result was not significant ($p = 0.200$), indicating that the distribution

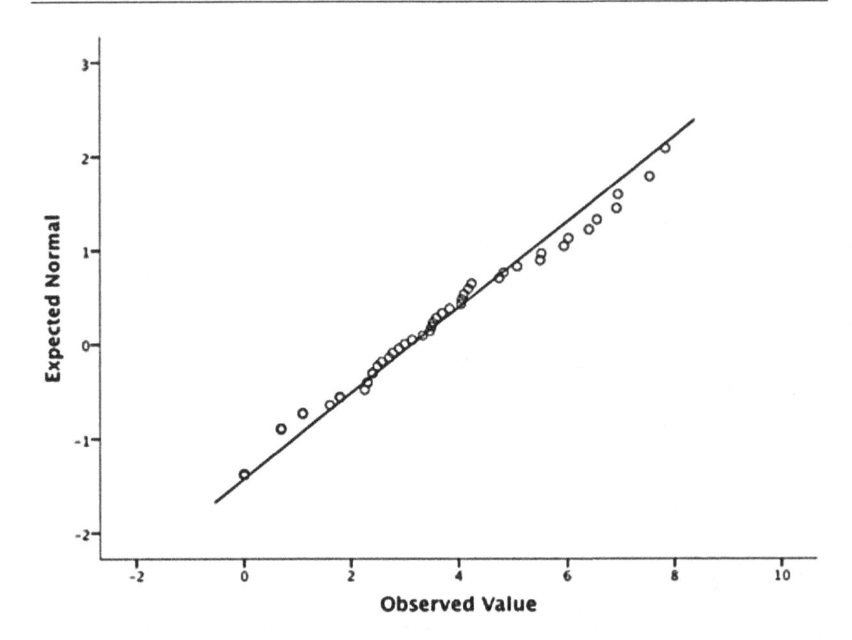

FIGURE 5.3 Q-Q Plot of Log-Transformed Cost [Reproduced with Permission of American Society of Mechanical Engineers, from "Impact of Advancing Technology on Nuclear Facility Operation," *ASCE-ASME Journal of Risk and Uncertainty in Engineering Systems, Part B: Mechanical Engineering*, Corrado, Jonathan and Ronald Sega, 6(1), 2020; Permission Conveyed through Copyright Clearance Center, Inc.].

of the log-transformed cost scores was not significantly different from a normal distribution. To test the second assumption of a *t*-test (homoscedasticity of variance), I used Levene's test to see if the variances of the groups were equal or unequal. The result was not significant ($p = 0.953$), so it can be assumed that the variances are equal, and the assumption was not violated. On average, the log-transformed costs of incidents determined not to be a result of human error related to technological advances were lower ($M = 0.92$, SD = 0.81) than the log-transformed costs of incidents determined to be caused by human error related to technological advances ($M = 1.99$, SD = 0.78), and this difference was significant, $t(51) = -4.81$, $p < 0.001$, one-tailed. The difference between the log-transformed means of cost between the two groups represented a large-sized effect, $d = 1.34$ [1]. Therefore, incidents caused by human error related to technological advances were more costly than incidents not meeting this criterion, thus rejecting the null hypothesis.

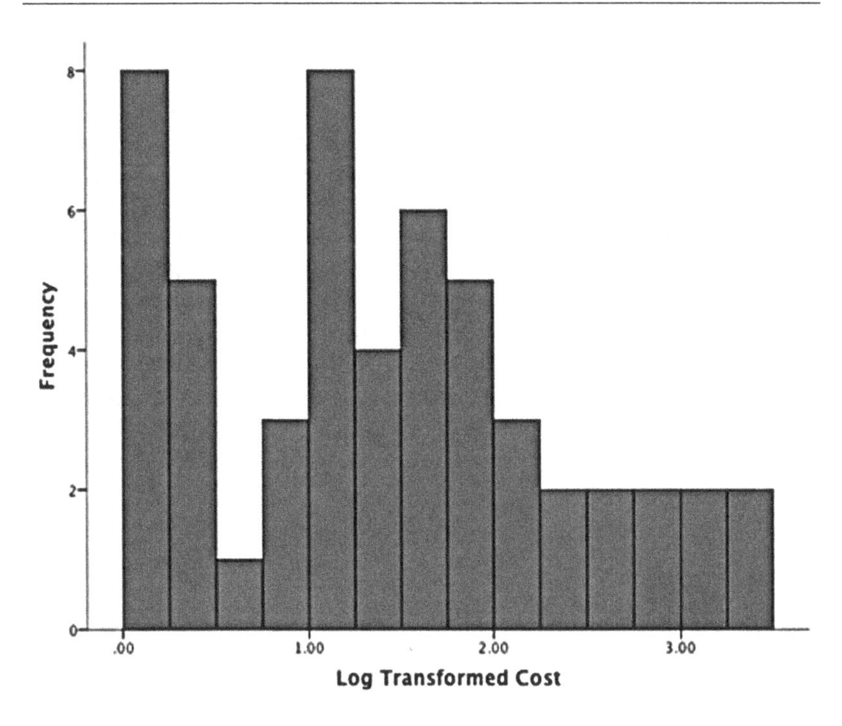

FIGURE 5.4 Histogram of Cost (Log-Transformed) [Reproduced with Permission of American Society of Mechanical Engineers, from "Impact of Advancing Technology on Nuclear Facility Operation," *ASCE-ASME Journal of Risk and Uncertainty in Engineering Systems, Part B: Mechanical Engineering*, Corrado, Jonathan and Ronald Sega, 6(1), 2020; Permission Conveyed through Copyright Clearance Center, Inc.].

In addition, I conducted non-parametric tests. The Wilcoxon-Mann-Whitney test is the non-parametric equivalent of the parametric independent samples *t*-test. The Wilcoxon-Mann-Whitney test does entail the assumptions that the data are normally distributed, the dependent variable is ordinal or continuous, and the independent variable consists of two categorical, independent groups. The present data meet these assumptions.

The results from the Wilcoxon-Mann-Whitney test suggested a statistically significant difference between the underlying distributions of the cost of incidents caused by human error related to technological advances and the cost of incidents determined not to be the result of human error related to technological advances, $z = 4.187, p < 0.001$. The average cost of incidents determined to be caused by human error related to technological advances was greater than the average cost of incidents determined not to be the result of human error related to technological advances.

Another possible way to deal with non-normally distributed data is to remove outliers. As seen in Figures 5.1 and 5.2, several outliers are present. Removing the four outliers with low costs ($2,483,000, $1,850,000, $1,025,000, $1,000,000) produced a somewhat more normal distribution (see Figure 5.5). However, the Kolmogorov-Smirnov test was significant ($p < 0.001$). Thus, the first assumption of the *t*-test was violated and the data with the outliers removed were not normally distributed. An independent samples *t*-test is not suggested when the data are not normally distributed. Excluded outliers are values that are approximately three standard deviations from the mean.

Because simply removing the outliers did not make the data normally distributed, the next alternative was to conduct a log-transformation of the data after removing the outliers. Log-transforming the data reduced the positive

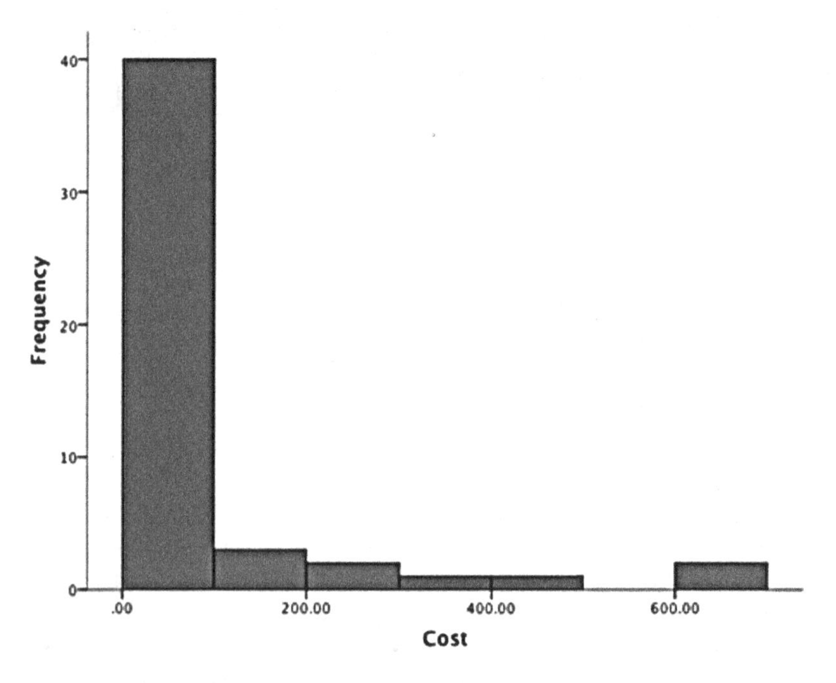

FIGURE 5.5 Histogram of Cost (Outliers Removed) [Reproduced with Permission of American Society of Mechanical Engineers, from "Impact of Advancing Technology on Nuclear Facility Operation," *ASCE-ASME Journal of Risk and Uncertainty in Engineering Systems, Part B: Mechanical Engineering*, Corrado, Jonathan and Ronald Sega, 6(1), 2020; Permission Conveyed through Copyright Clearance Center, Inc.].

skewness of the data, as seen in Figure 5.6. After the log-transformation, the data appeared more normally distributed (see Figure 5.7). Normality of the transformed data with no outliers was checked using the Kolmogorov-Smirnov test. The test was non-significant ($p = 0.200$), indicating that the distribution of the log-transformed cost scores was not significantly different from a normal distribution. I used Levene's test to check for homogeneity of variance; the result was non-significant ($p = 0.137$), so it can be assumed that the variances are equal and that the assumption was not violated. On average, the log-transformed costs of incidents determined not to be due to human error related to technological advances were lower ($M = 0.91$, SD = 0.81) than the log-transformed costs of incidents caused by human error related to technological advances ($M = 1.72$, SD = 0.59), and this difference was significant, $t(47) = -3.69$, $p = 0.001$. The results of the non-parametric tests again indicate

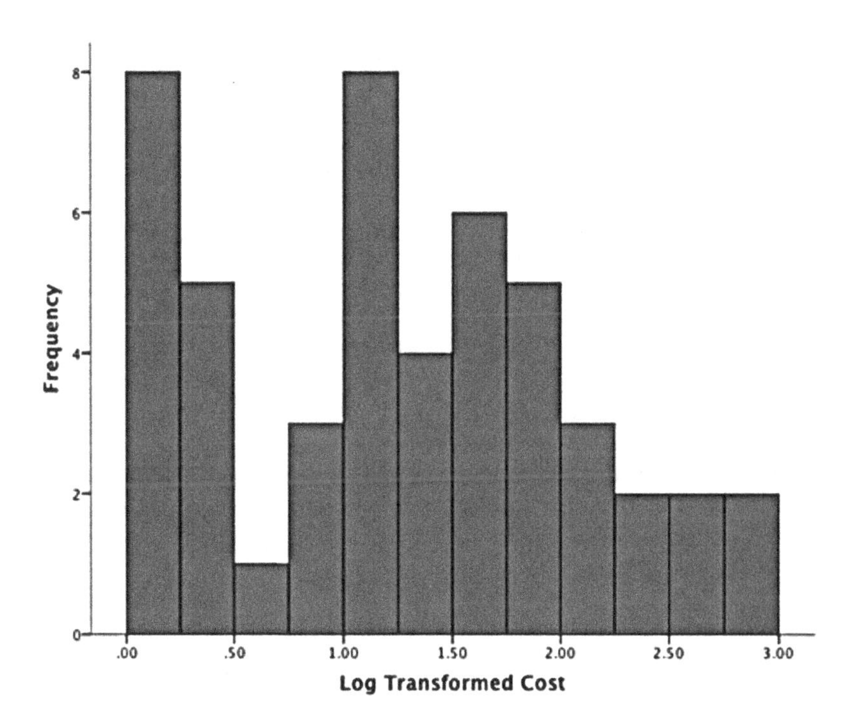

FIGURE 5.6 Histogram of Cost (Outliers Removed, Log-Transformed) [Reproduced with Permission of American Society of Mechanical Engineers, from "Impact of Advancing Technology on Nuclear Facility Operation," *ASCE-ASME Journal of Risk and Uncertainty in Engineering Systems, Part B: Mechanical Engineering*, Corrado, Jonathan and Ronald Sega, 6(1), 2020; Permission Conveyed through Copyright Clearance Center, Inc.].

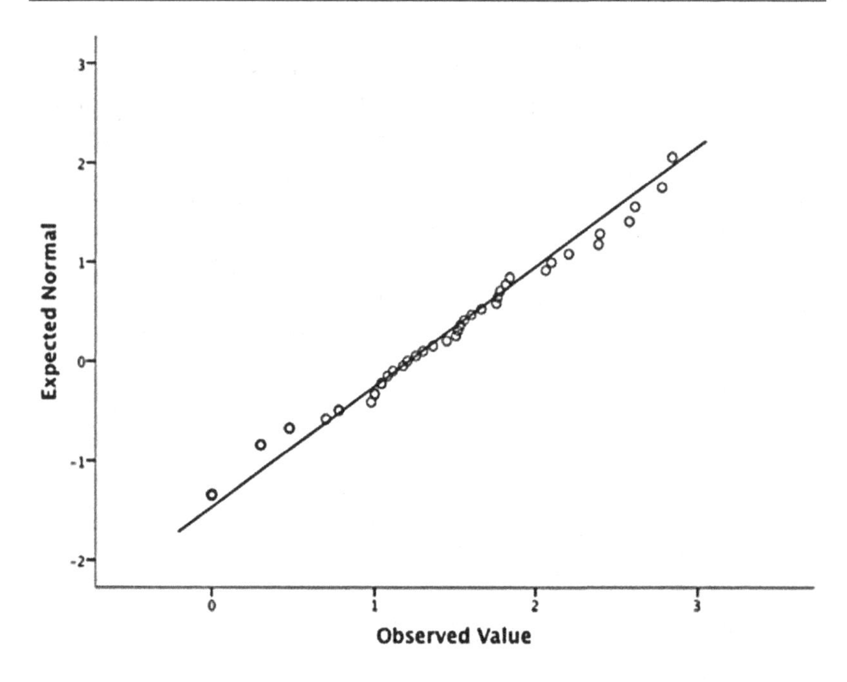

FIGURE 5.7 Q-Q Plot of Cost (Outliers Removed, Log-Transformed) [Reproduced with Permission of American Society of Mechanical Engineers, from "Impact of Advancing Technology on Nuclear Facility Operation," *ASCE-ASME Journal of Risk and Uncertainty in Engineering Systems, Part B: Mechanical Engineering*, Corrado, Jonathan and Ronald Sega, 6(1), 2020; Permission Conveyed through Copyright Clearance Center, Inc.].

that incidents determined to be caused by human error related to technological advances were more costly than incidents determined not to meet this criterion, thus further reinforcing the rejection of the null hypothesis.

INCIDENT FREQUENCY

As a means of further analysis regarding the frequency of incidents, the two groups of incidents were examined individually. A post-hoc *t*-test analysis was conducted for the twenty-two incidents in which human error due to technology advances was determined to be the cause.

I used the median-log-transformed cost (median = 4.12) to create two subgroups, low-cost and high-cost, within this group of incidents. An independent-samples *t*-test was conducted between the low-cost and high-cost subgroups. The analysis revealed a significant difference, $t(20) = -0.293$, $p < 0.008$. More interesting was the shape of the frequency distributions for high- and low-cost incidents. For lower-cost incidents, the frequency of occurrence appeared to approximate a normal distribution (Figure 5.8). The mean cost of a lower-cost technology-induced incident was $30.7 million (SD = $18.06 million). However, the high-cost group did not approximate a normal distribution (mean = $734 million; SD = $797 million). The frequency of incidents with costs over $500 million was much smaller. The relationship appears to be non-linear (see Figure 5.9).

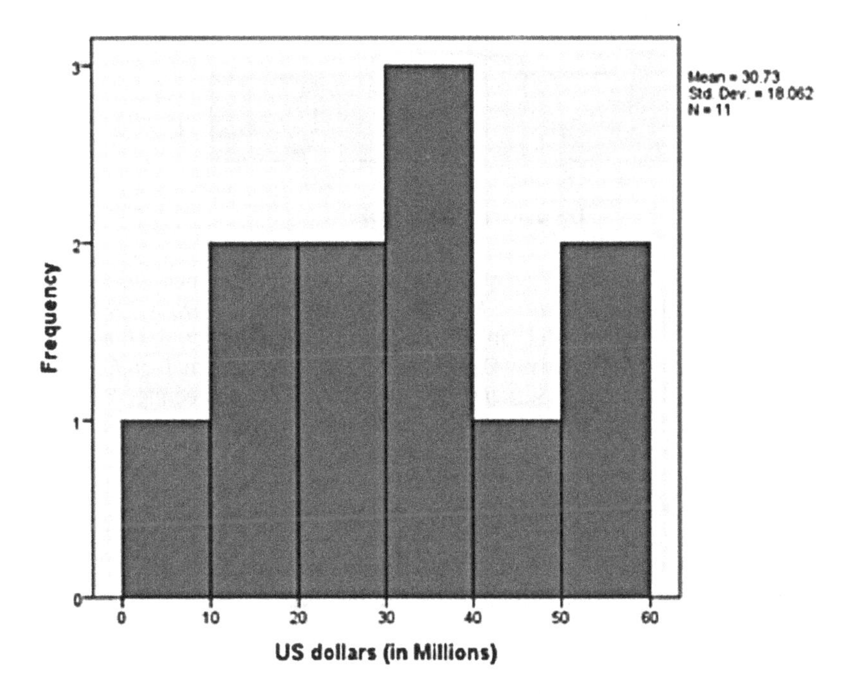

FIGURE 5.8 Histogram of Frequency Distribution of Cost per Incident for Lower-Cost Technology-Induced Human Errors [Reproduced with Permission of American Society of Mechanical Engineers, from "Impact of Advancing Technology on Nuclear Facility Operation," *ASCE-ASME Journal of Risk and Uncertainty in Engineering Systems, Part B: Mechanical Engineering*, Corrado, Jonathan and Ronald Sega, 6(1), 2020; Permission Conveyed through Copyright Clearance Center, Inc.].

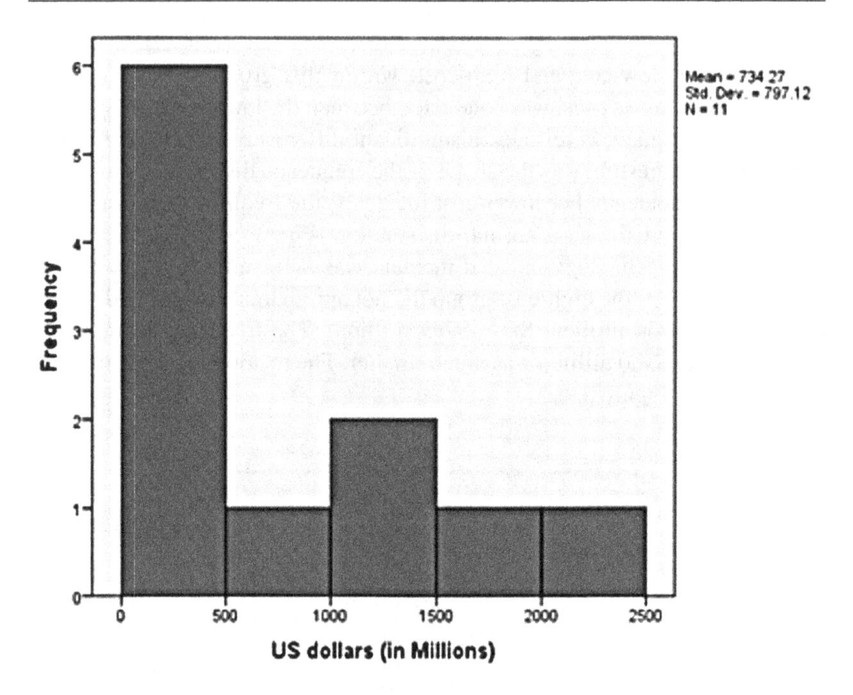

FIGURE 5.9 Histogram of Frequency Distribution of Cost per Incident for Higher-Cost Technology-Induced Human Errors [Reproduced with Permission of American Society of Mechanical Engineers, from "Impact of Advancing Technology on Nuclear Facility Operation," *ASCE-ASME Journal of Risk and Uncertainty in Engineering Systems, Part B: Mechanical Engineering*, Corrado, Jonathan and Ronald Sega, 6(1), 2020; Permission Conveyed through Copyright Clearance Center, Inc.].

To analyze the cost of incidents determined not to be the result of human error related to technology advances, a similar median split was performed on the log-transformed cost. An independent-samples t-test was conducted comparing the high- and low-cost errors in this group (n = 31). The analysis revealed a significant difference, $t(29) = -2.21$, $p < 0.035$. For lower-cost incidents determined not to be due to human error related to technological advances (n = 16), the frequency of occurrences did not approximate a normal distribution (Figure 5.10). The average cost for the lower-cost incidents determined not to be due to human error related to technological advances was $2.70 million. However, the cost per incident dramatically decreased after about $4 million. It is not that there are necessarily less incidents with a cost of $4 million or less. Rather, the cost per incident varies more and there is a greater gap in cost (i.e., some were way less than \$4 million) in low-cost incidents. Among the higher-cost incidents determined not to be due to human error related to technological advances ($n = 15$), again, the distribution did not approximate a normal distribution. The higher-cost incidents averaged \$109 million, but there were only three incidents at \$200 million or more (see Figure 5.11).

Finally, an independent samples t-test was conducted on lower-cost incidents, comparing incidents determined to be due to human error related to technological advances with incidents determined not to be due to human error related to technological advances. The analysis revealed a significant difference, $t(25) = 6.14$, $p < 0.0001$. Human error determined to be caused by technology advances had significantly greater cost impact (mean = \$30.73 million) than incidents determined not to be due to human error

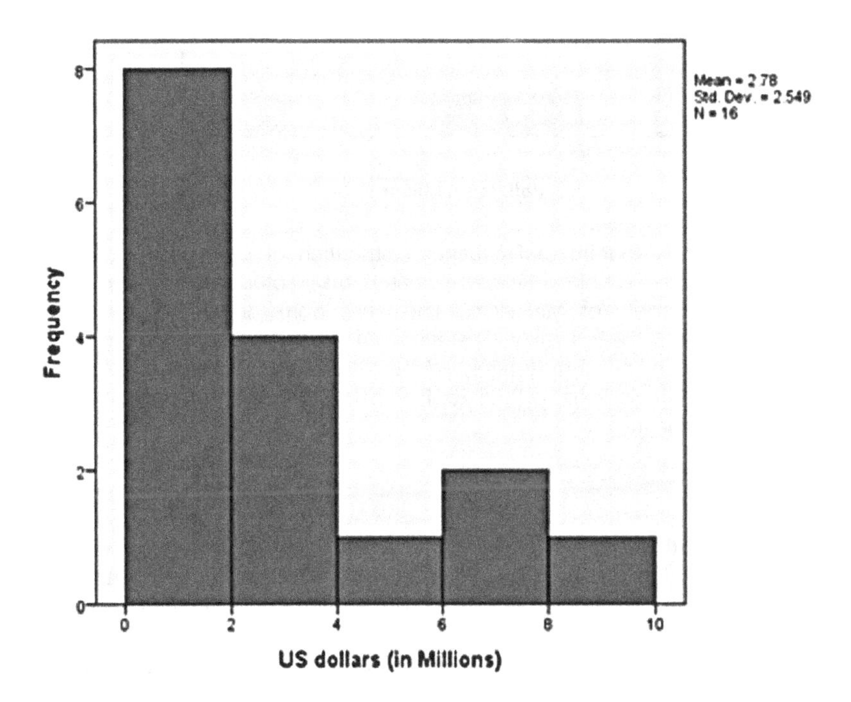

FIGURE 5.10 Histogram of Frequency Distribution of Cost per Incident for Lower-Cost Non-Technology-Related Human Errors [Reproduced with Permission of American Society of Mechanical Engineers, from "Impact of Advancing Technology on Nuclear Facility Operation," *ASCE-ASME Journal of Risk and Uncertainty in Engineering Systems, Part B: Mechanical Engineering*, Corrado, Jonathan and Ronald Sega, 6(1), 2020; Permission Conveyed through Copyright Clearance Center, Inc.].

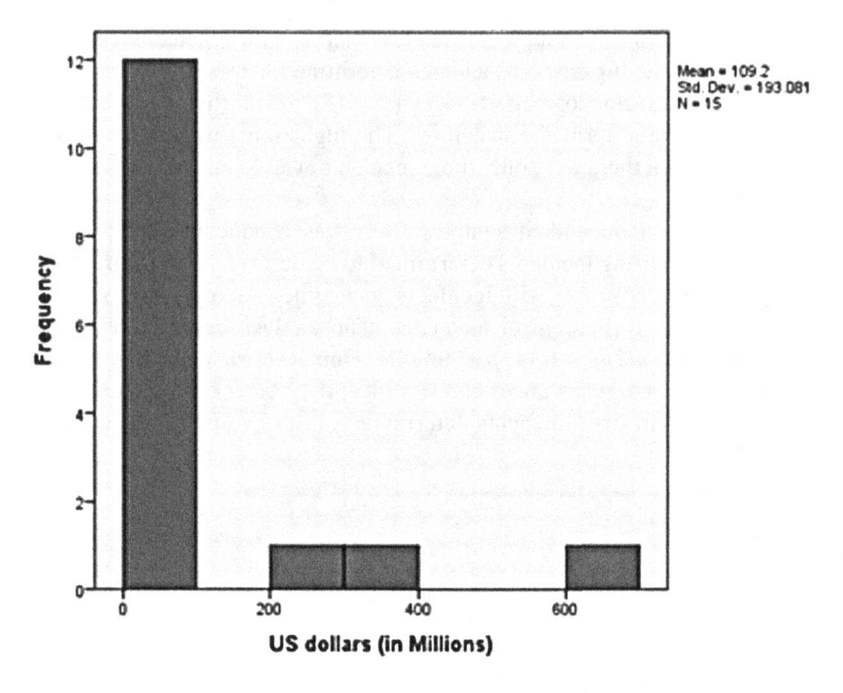

FIGURE 5.11 Histogram of Frequency Distribution of Cost per Incident for Higher-Cost Non-Technology-Related Human Errors [Reproduced with Permission of American Society of Mechanical Engineers, from "Impact of Advancing Technology on Nuclear Facility Operation," *ASCE-ASME Journal of Risk and Uncertainty in Engineering Systems, Part B: Mechanical Engineering*, Corrado, Jonathan and Ronald Sega, 6(1), 2020; Permission Conveyed through Copyright Clearance Center, Inc.].

related to technological advances (mean = $2.81 million) among "low-cost" incidents. In addition, an independent samples *t*-test revealed a significant difference between incidents resulting from human error related to technology advances (mean = $734 million) and incidents determined not to be due to human error related to technological advances (mean = $109 million), $t(24) = 2.94$, $p < 0.007$.

This additional analysis justifies the conclusion that incidents determined to be caused by human error as a result of technological advances are more costly than other incidents at nuclear facilities. A median split of the incidents determined to be due to human error related to technological advances found that lower-cost incidents (with a mean cost of about $30 million) were more evenly distributed around the mean than higher-cost incidents determined to be due to human error related to technological advances (mean cost of about $734 million). This finding confirms the above analysis, which indicated that incidents determined to be caused by human error related to technological advances are fewer in frequency, but more costly. Again, this results in a rejection of the null hypothesis.

THE RELATIONSHIP BETWEEN TECHNOLOGICAL IMPROVEMENTS AND ECONOMIC AND SAFETY FACTORS

This demands that we ask the question: Are organizations better off thanks to incorporating advanced technology at their facilities?

The Economics and Safety of Technological Improvements

To determine the economic and safety factors involved with technological improvements, information on twelve nuclear facilities for the years 2008–2013 was collected to represent a reasonable number of recent events. With regard to safety, I used the number of reported incidents as an independent variable and the capacity factor (explained in the next paragraph) as the independent variable for economic analysis. I used the cost of the upgrades as the dependent variable. An analysis of variance (ANOVA) and a regression analysis were performed on the data. I used the ANOVA to indicate the significance of the results; the regression equation indicated the extent to which the upgrades affected the plant's capacity factor and the number of reported incidents [1].[1]

The capacity factor for a power plant is calculated as the ratio between the observed output over a given period of time and the potential output if the facility were consistently running at its nameplate capacity. The calculation of the capacity factor involves dividing the amount of energy produced by a plant during a certain time by the amount of energy that would have been produced if the facility were always running at full capacity. The facility's capacity factor depends on the type of fuel used, the facility's design, and its age, among other factors.

Interpretation of the Outcomes

An examination of Figures 5.12 and 5.13 and consideration of the regression coefficients reveal that increasing the number of upgrades is associated with a higher capacity factor (a positive outcome) and more incidents being reported (a negative outcome).

The US Energy Information Administration calculated the capacity factors for various types of power plants in the United States in 2009. Oil plants had the lowest capacity factor at 7.8 percent. Other types of power-generating facilities had higher capacity factors, with hydroelectric at 39.8 percent, other renewables at 33.9 percent, natural gas at 42.5 percent, and coal at 63.8 percent. In contrast, the capacity factor for US nuclear facilities during 2009 was 90.3 percent. The capacity factors for the twelve facilities examined in this study from 2008 until 2013 ranged from 70 percent to 113 percent. Capacity factors in excess of 100 percent can be achieved by system upgrades that result in capacity's exceeding the initial plant design capacity. Compared to other forms of fuel, nuclear power has a distinct advantage with regard to its capacity factor [2].

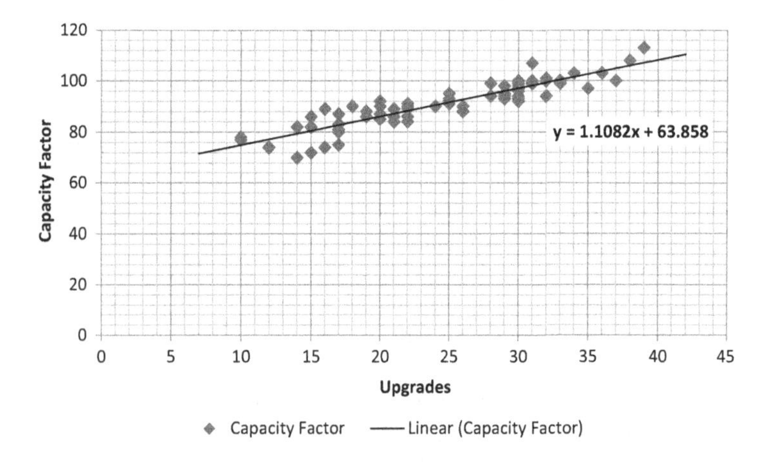

FIGURE 5.12 Scatterplot Showing the Relationship between Upgrades and Capacity Factor [Reproduced with Permission of American Society of Mechanical Engineers, from "Impact of Advancing Technology on Nuclear Facility Operation," *ASCE-ASME Journal of Risk and Uncertainty in Engineering Systems, Part B: Mechanical Engineering*, Corrado, Jonathan and Ronald Sega, 6(1), 2020; Permission Conveyed through Copyright Clearance Center, Inc.].

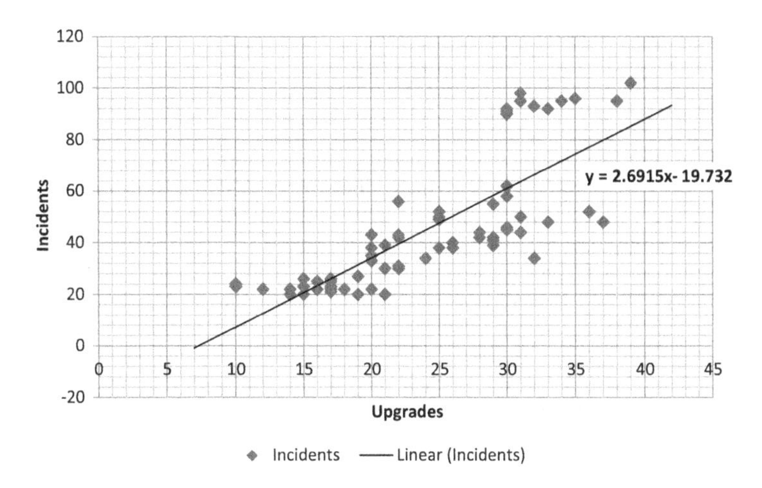

FIGURE 5.13 Scatterplot Showing the Relationship between Upgrades and Incidents [Reproduced with Permission of American Society of Mechanical Engineers, from "Impact of Advancing Technology on Nuclear Facility Operation," *ASCE-ASME Journal of Risk and Uncertainty in Engineering Systems, Part B: Mechanical Engineering*, Corrado, Jonathan and Ronald Sega, 6(1), 2020; Permission Conveyed through Copyright Clearance Center, Inc.].

The number of incidents was recorded as a way of measuring the level of safety associated with the upgrades. As the amount of money spent on upgrades increased, the complexity of the system rose as well as the number of incidents. However, these were only incidents which were at a severity Level of 1 or 2. None of the twelve facilities randomly chosen for this specific analysis had a major incident during the six years of data collected [1].

The data indicate that spending more money on upgrades can increase the capacity of the facility as well as the potential number of incidents.[2] However, the incidents that occurred at these twelve facilities over the six years studied were relatively minor. Given that nuclear facilities produce vast amounts of power and that the upgrades significantly increased the capacity factor, there appears to be a financial advantage in conducting upgrades; however, this should be weighed against the increased rate of Levels 1 and 2 incidents observed. Again, many incidents at Level 3 or higher appear to have been associated with human error while working with new technological advances. Because it does not seem likely that nuclear power facilities can become fully automated at any point in the near future, engineers must continue to increase facility capacity factors with upgrades while minimizing human errors [1].

NOTES

1 There were twelve facilities in the sample and the information was collected over six years, yielding seventy-two data points for the ANOVA, which implies seventy-one degrees of freedom. The F-test yielded an F statistic of 208.86, which was significant at the $p < 0.01$ level. The regression analysis indicated that the regression coefficient for capacity factor was 0.592. The coefficient for incidents was 0.075. Both of these coefficients indicate that an increased number of upgrades would result in a higher capacity factor for the facility. However, the number of incidents also increases with greater investment in upgrades [1].

2 For more, see Mark L. Mitchell, and Janina M. Jolley, *Research Design Explained*, 8th ed. (Belmont, KY: Wadsworth Cengage Learning, 2013).

REFERENCES

1. Republished with permission of American Society of Mechanical Engineers, from Corrado, Jonathan, and Ronald Sega. 2020. "Impact of Advancing Technology on Nuclear Facility Operation." *ASCE-ASME Journal of Risk and Uncertainty in Engineering Systems, Part B: Mechanical Engineering* 6(1). doi:10.1115/1.4044784; permission conveyed through Copyright Clearance Center, Inc.
2. U.S. Energy Information Administration. 2011. *Electric Power Annual 2009* (DOE/EIA-0348). Washington, DC: U.S. Department of Energy.

6 The Application of the Systems Engineering Process to Enhance Human Performance Improvement in the Operation of Nuclear Facilities

THE IMPACT OF HUMAN FACTORS ON PLANT OPERATION

To recognize the design considerations necessary to combat human error during system design and integration, we must first understand the impact of human factors on the complexities of plant operations. Both the human mind and the systems with which it interacts at these facilities represent high levels of sophistication often masked as advanced technology becomes increasingly incorporated into system design. However, progress has been made toward determining the root causes of

DOI: 10.1201/9781003346265-6

these problems. Detailed investigation of human errors indicates that both micro- and macro-ergonomic causes. The incidents are nearly always the result, in some form, of how the systems operate and humans interact with them [1].

As the systems themselves are designed by humans, and errors can also occur in the design phase. Designers attempt to create systems so that operators can work effectively with them and avoid errors. However, an error at the design level can ultimately result in an operator error. Moreover, many errors are attributable to other aspects of the complicated operational processes involved, such as unresponsive management systems, ineffective training, organizational designs that are not adaptable, poorly designed response systems, and environmental disturbances of a chronic nature [2].

HUMAN BEHAVIOR MODELS

Ergonomics

Understanding human error requires understanding human behavior. Human factors engineering, or ergonomics, is concerned with improving the safety, health, productivity, and comfort of workers. It is also concerned with enabling smooth interaction between the people using the technology and the environment within which they are working, often called the human-machine environment. When this context is considered at the micro-ergonomic level, the focus is on the level of the human and machine, such as the design of the individual control panels, workstations, the visual displays, and the ergonomically fitted seats on which employees such as nuclear facility operators spend most of their work hours sitting. Problems that arise at this level may be due to improperly designed displays and workstations [1]. For example, many problems identified in the analysis of nuclear incidents resulted from the installation of new technology that had not been designed well in ergonomic terms. In one incident, the operator reported that the display was too bright and that after three hours on the job, the operator could no longer view the display clearly. This difficulty caused the operator not to notice that the power plant was operating outside its normal parameters.[1]

Also, within the realm of human factors and ergonomics are considerations related to the operators' body size, or anthropometrics, as well as human decision-making, cognitive capacity, information processing, and overall human skills [3]. One real-life impact of anthropometrics involved an incident that occurred when an operator was away from his station. Normally, operators remained at their station with only a few short breaks during an eight-hour shift.

However, new seating had recently been installed. It was supposed to improve comfort as the company that provided the seating had a long history with the nuclear industry and good reviews by nuclear plant operators. Indeed, most operators rated the seating as excellent with regard to comfort. However, one particular operator was unusually tall at 6′7″ and experienced difficulty with the new seats designed for shorter people. As a result, the operator began taking more frequent and longer breaks to stretch and avoid muscle cramps. While the operator was away from his station, a release valve malfunctioned, and the system continued to operate longer than should have been permitted given the state of the valve [2].

The building blocks for technological systems such as nuclear facilities include personnel and engineered components. However, the organization and its structure can also be important. These aspects of the workplace are referred to as macro-ergonomics. In both of the cases just described, the operators had reported the problem to their supervisor. The operator with the bright display panel had reported the problem once; the one with the uncomfortable seat had complained twice verbally and once in writing prior to the incident. If either of these supervisors had followed up on these complaints, an adverse incident may have been avoided. Therefore, it is crucial to have systems in place that encourage operators working with new equipment to alert their supervisor if difficulties arise. With regard to human factors, usually, only the operators of the facility can gauge the success of the new system. Often, problems related to human factors engineering are not evident until a complaint or concern is raised, or an incident occurs [1].

Performance levels and the inherent potential for mishaps in complicated technological systems are usually a function of the human and engineered subsystems. The engineering may include such items as workstation design and the appearance of control boards; human engineering encompasses organizational and personnel systems as well as operational matters. Many system failures (typically about 70 percent at nuclear facilities) are attributed to operator error, but this is often an oversimplification because it does not fully take into account various factors beyond the control of the operator, such as ineffective response systems, organizational designs that are not adaptable, unresponsive management systems, ineffective training, and overly complicated operational processes. These threats can be especially problematic when new equipment has been installed [1].

THE IMPACT OF HUMAN ERROR ON PLANT OPERATION

Reducing the negative impact of technological advances on human error entails improving human performance. This type of performance improvement is a systematic process that seeks to analyze and discover performance gaps, monitor performance, determine the desired level of performance, and develop effective interventions. Once the interventions have been developed, they must be implemented and continually evaluated with regard to their results. The ultimate goal for human performance is to make the nuclear facility as close to event-free as possible. This can be achieved through the proactive management of human performance. It is also necessary to strengthen the facility defenses along with operator performance. Optimization of both the organization and operating processes can reduce errors to a minimum level [1, 4].

Even the best-trained and most motivated operators of nuclear facilities will still make mistakes. No amount of training or coaching can prevent all errors, as the interaction between the organization, the workplace, work tasks, operating systems, and operators creates the potential for a wide range of errors. The first step in preventing further problems is to understand why and how problems occur [1, 4].

Errors at a nuclear facility can be reduced through using self-checks and specific tools. Random errors can never be fully eliminated, but they can be reduced. For example, a maintenance professional at the facility may be required to tighten a valve. No matter how well trained the individual is, there is always a chance of a mistake when this task is performed. The valve may be too tight or not tight enough. The question to ask is why the individual has made the mistake and what can be done to prevent future occurrences. In the case of such a mistake, multiple barriers should be present to prevent system failure, so as to minimize the possibility of severe consequences. There must also be an organizational infrastructure in place that both identifies errors and protects the operators from injury or death. This creates a situation in which more significant problems can be avoided [1].

Although errors are inevitable, steps can be taken to manage, predict, and prevent them. One beneficial approach is to recognize error traps and communicate them to others so as to proactively manage problem situations and minimize errors. The work situation can be changed to reduce, prevent, or remove conditions that lead to errors. Individual factors and tasks can be altered within the work environment to minimize future errors [1]. For example, in the case of the seats that made an extremely tall operator uncomfortable, the difficulty could have been prevented if the manufacturer had requested a physical description of the operators [5].

Organizational values and processes significantly influence the behavior of an individual working at a nuclear facility. The values and processes of an organization can be developed in a manner that encourages individuals to take actions that increase the chances of achieving the organization's goals. In the case of a nuclear plant, the organization's values may focus on precision, accountability, and excellence for all individuals working at the facility. This would encourage safe behaviors that decrease the number and severity of errors. Facility managers can guide workers' behavior toward producing results that are more desirable and contain fewer errors. Improving staff performance requires excellence with regard to management systems, culture, and organizational processes. Exploiting the social interactions involved in work at a nuclear facility in positive ways can significantly decrease the likelihood of errors [1].

Improvements in human performance can be achieved through taking corrective actions after analyzing problem reports and events. These corrective actions are part of the learning that should ensue after a problem occurs. Though reactive rather than proactive, they are important for improving the systems and technology involved. Combining reactive and proactive methods of learning facilitates anticipation of problematic events and the prevention of errors. This is often more cost-effective than using only the reactive approach [1].

The collective behaviors of people at all levels in a facility determine the performance outcome achieved. The individual's work is a product of their mental processes, which have been influenced by a variety of factors and demands which are present in the work environment. The work is also a function of the capabilities of each of the people involved. When the facility achieves high performance, its individuals will usually take responsibility for their own behaviors. They will also be committed to personal improvement as well as to improving the work environment and their completion of tasks. Individuals working in a high-performance facility will be active in confirming the integrity of the facility defenses, anticipating situations likely to precipitate errors, and communicating with others to create a shared understanding of the facility and the work to be done [1].

THE IMPACT OF HUMAN ERROR ON THE SYSTEMS ENGINEERING PROCESS

To understand how human error can be manifested in system design, we must grasp the key aspects of the systems engineering process and the complexity and potential for human error that are inherently imbedded [4].

Technological and other systems are increasingly complex and challenging. Simpler systems, especially those related to information, are typically old, incapable of meeting current demand, and rapidly becoming obsolete. To facilitate system upkeep, safety, and reliability and to ensure that innovation is consistent and ongoing to meet clients' needs, the systems engineering process must be relatively complex [4].

The systems engineering process is composed of several progressive stages and encompasses a breadth of evaluations and considerations that are performed as a function of the process. To comprehend how human performance impacts this process and vice versa, a brief discussion of the process is necessary [4].

Although there is common agreement regarding the doctrines and intentions of systems engineering, execution differs from one system or design team to the next. The steps involved and the general methodology will depend on the background and prior experiences of the individuals on the design team. The most common and widely accepted systems engineering process and life-cycle progression are displayed in Figure 6.1 [4].

The process begins with the identification of a need, such as to replace an obsolete system with a serviceable one, to improve a system in order to increase plant efficiency or capacity, or to upgrade a system due to new regulatory requirements. Once the need has been identified, then the conceptual design stage commences with the creation of a project management plan. During this stage, the problem is formally defined, specific system needs are identified, requirements are analyzed, maintenance and support are conceptualized, technology is evaluated, and the technical approach is selected. Conceptual design focuses on the system as a whole. At the conclusion of conceptual design, the systems engineering management plan (SEMP) and the test and evaluation master plan (TEMP) are written, the conceptual design review (system requirements review) is conducted, and the functional baseline system specification is defined [4].

Next comes the preliminary design stage. During this stage, a functional analysis is conducted, requirements allocation is performed, trade-off studies are executed, early prototyping may be achieved, and acquisition, contracting, and supplier activities are accomplished. This stage focuses on the subsystem level. At the conclusion of this stage, the system design review is conducted,

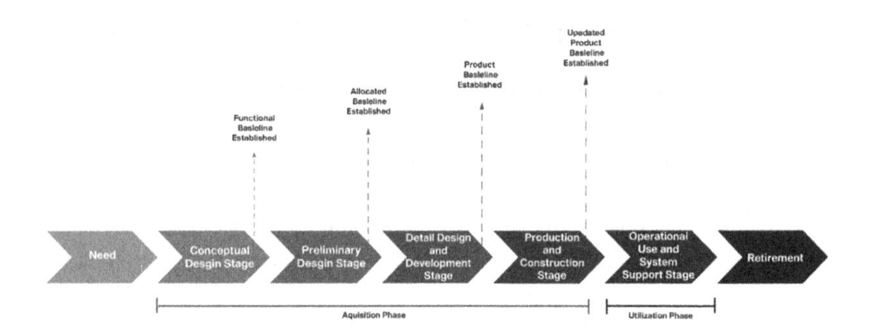

FIGURE 6.1 The Systems Engineering Process [Modified with Permission of American Society of Mechanical Engineers, from "Proactive Human Error Reduction Using the Systems Engineering Process," *ASME Journal of Nuclear Engineering and Radiation Science*, Corrado, Jonathan, 8(2), 2022; Permission Conveyed through Copyright Clearance Center, Inc.]

and the allocated baseline, development, process, product, and material specifications are defined [4].

After preliminary design comes detail design and development. This stage consists of subsystem/component design, additional trade-off studies and evaluation of alternatives, development of engineering and prototype models, and development test and evaluation. This stage focuses on the component level. At the conclusion of this stage, the critical design review takes place, and the product baseline, process/product, and material specifications are defined [4].

Next comes the production/construction stage. This stage consists of the production and/or construction of system components, acceptance testing, system distribution and operation, development and operation test and evaluation, and system assessment. This stage focuses on modifications for improvement of the designed system. At the conclusion of this stage, an updated product baseline is produced [4].

The last stage in the systems engineering process is the operational use and system support stage. This stage consists of system operation in the user environment, sustaining maintenance and logistic support, operational testing, system modifications for improvement, contractor support, and additional system assessment (field data collection and analysis). This stage again focuses on modifications for improvement. At its conclusion, there exists an operating system that adequately fulfills its original intended purpose, meets acceptance criteria for all tests and milestones, and operates until its eventual retirement and disposal [4].

Feedback is integral to the systems engineering process. It is important that the right information be reported and fed back to the responsible engineering and management personnel promptly and efficiently. Those responsible need to know in a timely fashion exactly how the system is performing against specifications in the field, so that design modifications can be initiated. The primary objective is to provide a good assessment of just how well the system is performing in the user's operational environment; a secondary objective is to identify any problems and initiate the required steps leading to corrective actions and the incorporation of necessary design changes and system modifications [6].

ISSUES RELATED TO ADVANCED TECHNOLOGY AND THE TWENTY-FIRST-CENTURY OPERATOR

To this end, today's sensory and processing technologies are perceptive and precise. They can discern the environment, solve complicated problems, make assessments, and learn from experience. Although they do not think as humans do, they can replicate many human intellectual aptitudes. Throughout the last century, for varying reasons, companies have implemented advanced technology and removed human from many aspects of operation [1].

Human reliance on technology may be demanding a high price. Is our own understanding declining as we become more dependent on advancing technology and its scale of influence broadens? Computers have ventured into many different kinds of knowledge work: pilots rely on computers to fly planes, doctors consult them in diagnosing illnesses, and architects use them to design buildings. Automation is impacting virtually every industry. Computers are not taking away all the jobs performed by talented people, but they are changing how the work gets accomplished.

As technology continues to develop, the people using it become less likely to refine their own capabilities. Technology that offers many prompts and tips could be responsible for this trend; simpler, less helpful programs push operators to think harder, perform, and learn. Humans' skills become sharper only through practice when we use those skills regularly to overcome different and difficult challenges.

There should also be a balance between advancing technologies integrated into plants and the interaction between humans and these new technologies. New technologies should be designed and incorporated into existing systems to keep the human operator in the decision cycle, which consists of an ongoing process of action, feedback, and judgment. This will ensure that operators remain attentive and engaged and promotes the kind of challenging activity that strengthens skills. Technology should play an essential but secondary role. It should broadcast warnings when parameters are exceeded, provide vital information that enhances the operator's outlook, and protect against the biases that often alter human thinking. The technology will then become the operator's partner, not the operator's replacement. This approach to technology application will not stifle technological progress. It requires only a slight shift in design priorities and a rekindled emphasis on human strengths and weaknesses.

Incorporating advances in technology is, in many regards, a necessary enterprise, but should be properly balanced within the confines of the system. It is easy to forget that humans are a vital part of this system. Technology is strongly suited to perform many functions, but it lacks the ability to rationalize. Decisions concerning the incorporation of new technology into a plant must consider the human factor. Even the smartest software lacks the common sense, ingenuity, and vitality of the skilled operator. In the control room, human experts remain indispensable. Human insight, imagination, and perception, enhanced through hard work and experience, cannot be replicated by the most cutting-edge technologies today and into the near future. If we let our own skills fade by relying too much on the technology crutch, we will render ourselves less capable, less resilient and more submissive to our machines.

Cognizance of these realities is imperative during system design. The person, including the human's propensity for error, should be considered a vital element of the system considered in design and accounted for through a rigorous systems engineering process. Engaged human participation is compulsory for successful system operation, but like all systems, it has its failure modes. People's natural susceptibility for error in system operation should be addressed from multiple fronts.

THE IMPROVEMENT OR UPGRADE OF EXISTING SYSTEMS

To aid this, many organizations improve existing systems toward three purposes: replacing obsolete technology or equipment, regulatory reasons, and economic motives. Therefore, the outcomes of the improvement or upgrade of existing systems can be observed as two-pronged, encompassing safety-related and economic factors.

A Way Forward Leveraging the Systems Engineering Process

It is odd, then, that human error reduction and system design and deployment are often treated as two separate subjects with their own distinct processes that commonly intersect upon the conclusion of design, immediately prior to operation. This traditional approach to system design may have been acceptable for the antiquated, obsolete technologies of the past, but it is problematic for designing today's more complicated systems. As the complexity of advancing technologies crescendos, human-system interaction warrants a more prominent role in system design and therefore compels early consideration, deliberation, and integration in the beginning stages of the systems engineering process. Incorporation of methods of human error prevention into the systems engineering process yields the fruit of the sound development of systems with an improved probability of successful operation and reduced error frequency [4, 7].

At this juncture, it is important to discuss the difference between human performance and human factors as they relate to system design. I discuss both concepts throughout this book, but there are subtle differences between them. Human factors in the context of the study of design refer to the practice of designing products, systems, or processes to take proper account of the interaction between them and the people who use them. In substance, it is the study of designing equipment and devices that fit the human body and its cognitive abilities. Alternatively, human factors in the context of the study of humans, can be described as the study of how human beings function within various work environments as they interact with equipment in the performance of various roles and tasks (at the human-machine interface). Common human factors considerations include anthropometric, sensory, physiological, and psychological factors. Human factors, like reliability and maintainability, are an inherent consideration within and throughout the systems engineering process [7].

Meanwhile, human performance is a field of study related to process improvement methodologies that can reduce human errors. It focuses on improving performance at the societal, organizational, process, and individual levels. In other words, it is a series of behaviors executed to accomplish specific results. Human performance improvement (HPI) methodologies are traditionally initiated after systems or plants are designed and installed and plant operations are underway; they are not conventionally employed, at least explicitly, during system design [7].

In essence, human factors influence human performance, but human performance does not necessarily influence human factors in the design of a system. Human performance, not human factors, is the primary concept in view in this chapter [7].

As shown in Chapter 5, human errors resulting in incidents at nuclear facilities can be very expensive and potentially harmful to the employees, the public, and the environment. Also in Chapter 5, it was found that incorporating advanced technologies into nuclear facilities does increase capacity, but also results in an increased frequency of incidents. These incidents affect the public's perception of the company involved and the nuclear industry generally, along with stock values, insurance rates, regulatory oversight, and potential civil penalties. Following incidents at nuclear plants, operating experience and human performance lessons have been documented, disseminated to the industry, and incorporated into plant operations to ensure that similar situations would not occur in the future. The incorporation of industry operating experience and human performance methodologies into plant operation is a routine and expected aspect of nuclear plant operational life that promotes continuous improvement and a constant striving for safe operations. If human performance considerations have a positive impact after deployment, then it would be logical to consider that they could have a positive impact if taken into account before deployment. The focus of this chapter is the systematic incorporation of human performance into all stages of the systems engineering process. Figure 6.2 depicts a proposed integration of HPI throughout the systems engineering process [4, 7].

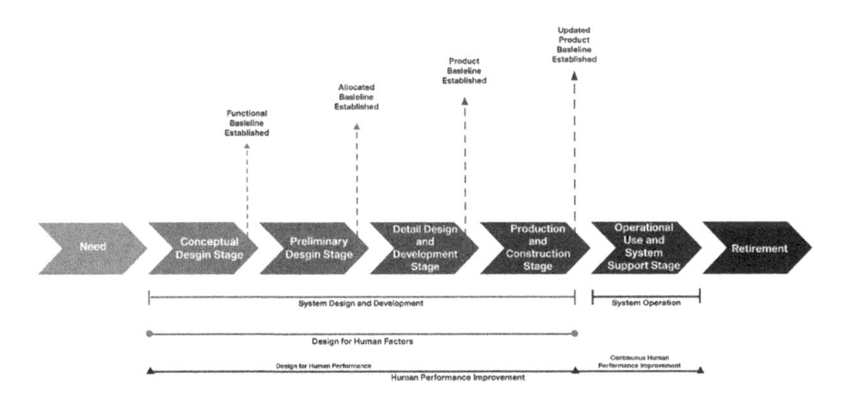

FIGURE 6.2 Human Performance Improvement throughout the Systems Engineering Process [Modified with Permission of American Society of Mechanical Engineers, from "Proactive Human Error Reduction Using the Systems Engineering Process," *ASME Journal of Nuclear Engineering and Radiation Science*, Corrado, Jonathan, 8(2), 2022; Permission Conveyed through Copyright Clearance Center, Inc.]

The cultivation of human performance–enhanced system design and operation can stem from (1) operator involvement in the systems engineering process; (2) human performance association with system operational requirements and system testing, evaluation, and validation; (3) iterative procedures and operator training development throughout all stages of the systems engineering process; and (4) selection and cultivation of aptly inclined operators chosen and groomed specifically for the systems being designed. Figure 6.3 illustrates the proposed human performance interface with the systems engineering process [4]. As shown, feedback exists throughout the process, not only between the steps of the traditional systems engineering process, but between the systems engineering process and the human performance elements. The human performance feedback is represented as a solid line to articulate a more focused role in this context, whereas feedback in the traditional process is represented as a dashed line as it is more general in nature. Feedback goes both directions in all cases, as the process iteratively progresses.

Operator Involvement in the Systems Engineering Process

Whether a system is being newly created or an existing system is undergoing a minor modification, plant operations personnel should be involved in the systems engineering process from the onset. Operators are a significant system stakeholder.

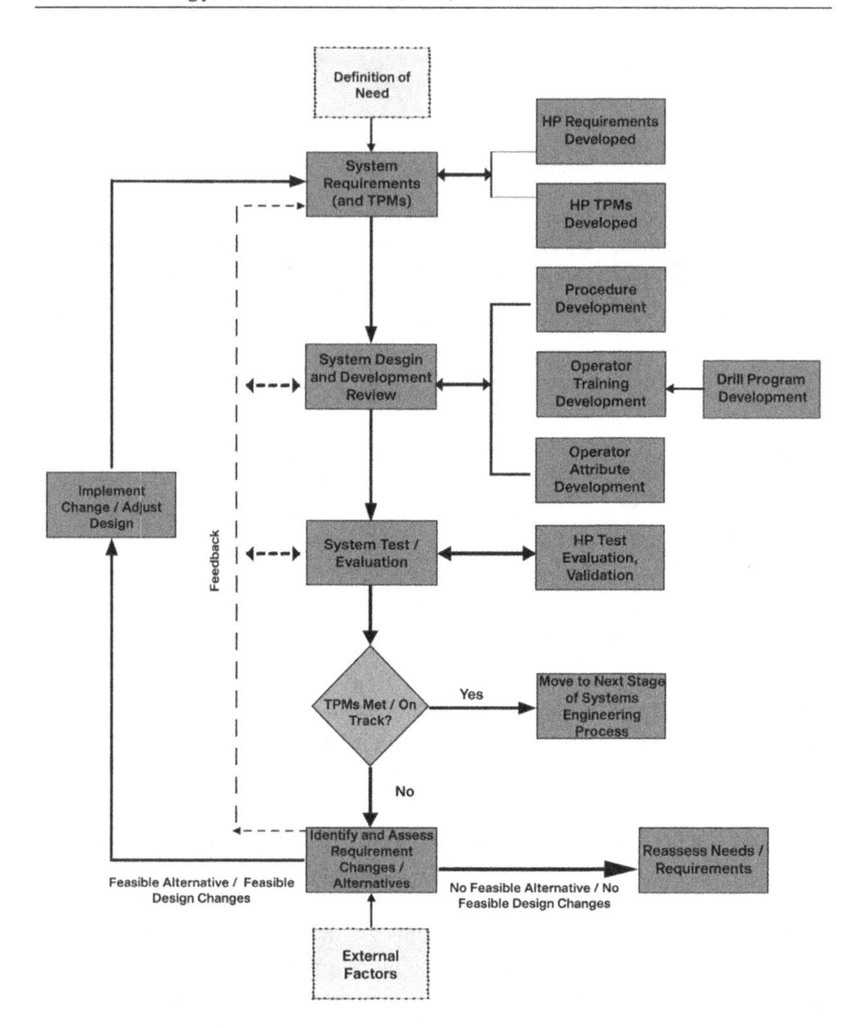

FIGURE 6.3 Human Performance Interface with the Systems Engineering Process [Modified with Permission of American Society of Mechanical Engineers, from "Proactive Human Error Reduction Using the Systems Engineering Process," *ASME Journal of Nuclear Engineering and Radiation Science*, Corrado, Jonathan, 8(2), 2022; Permission Conveyed through Copyright Clearance Center, Inc.]

Not only do they bring a unique yet vital perspective to the design team, but they will eventually inherit the system being designed. Operators bring an essential perspective from the field, understanding the environment in which the system will operate, how the operator will interface with the system, and the robustness and redundancy that the system will need to possess to operate as required [4].

Achieving balance of innovation and practicality is often difficult, as designs are conceptualized, iterated, and deployed. The process may necessitate reverting to the design stages for necessary alterations. In most cases, some duplication of effort can be avoided by appointing members of the operations staff to the system design team and leveraging their experience from initial identification of needs through system design and development [4].

Lifecycle Engineering

Designing for the system lifecycle is a foundational concept of systems engineering. The system lifecycle begins with the needs identification process and continues through design and acquisition to usage. Lifecycle engineering transcends the siloed, isolated view of systems by embracing all aspects of the system including maintenance, support, operation, and eventually phase-out and disposal. Because system use encompasses a large portion of the system's life, the participation of operations personnel in system design is vital not only for system operational success and support, but also for human error reduction in the operation of new technologies.

Failure Mode and Effects Analysis

To aid with knowledge of potential errors, failure mode and effects analysis (FMEA) is a design method that can be used to identify and examine potential system weaknesses. It includes the necessary steps for examining all the means by which a system failure can occur, the potential consequences of failure for system performance and safety, and the significance of these effects [8]. FMEA can be used during all stages of the systems engineering process. Operations personnel can play a role in FMEA during all stages of system design. Having operated similar systems, having been exposed to a breadth of systems failures, and having witnessed and responded to the resulting impact of these failures on the plant, operators have valuable insights that should be incorporated throughout design and development.

System Design Evaluation

The fulcrum of the systems engineering process lies in system design evaluation. Evaluation is an obligation intrinsic to the systems engineering process and must be conducted frequently as the design activities progress. Likewise, the establishment of a clear set of system design criteria must precede evaluation. Design criteria can and should come from a number of diverse sources, but those that originate from the plant operations staff should be given elevated attention because the operators will in effect be the primary system owners throughout the system's useful life.

Evaluation in the context of systems design requires finesse and fidelity. System design should harmonize union, examination, and evaluation, and these technical activities should be integrated and employed iteratively and persistently over the system lifecycle. The active engagement of operations personnel is central to all these steps.

The Association of Human Performance with System Operational Requirements and System Testing, Evaluation, and Validation

The conceptual design stage is the first and most important stage of system design and development because it lays the groundwork for what follows. As discussed earlier in this chapter, the identification and definition of a need is required to provide an effective and proper starting point for this stage. Inherent to this foundational stage are the generation of operational requirements and the establishment and commencement of system testing, evaluation, and validation [4].

Human Performance and System Operational Requirements

Upon designation of the need and technical approach for the system, the operational requirements are then defined. Because system operational requirements should be identified and specified early, wisely, and as completely as possible, the identification of human performance requirements should occur at this point in the process [4].

System operational requirements address the following elements: mission definition, performance, and physical parameters, operational deployment, operational lifecycle, utilization requirements, effectiveness factors, environmental factors, and economic factors. Embedded in these elements should be specific, exclusive human performance system requirements that can be easily assessed and discernible in system design. Once human performance system requirements have been established, specific human performance technical performance measures (TPMs) can be detailed [4].

Human Performance and System Testing, Evaluation, and Validation

System testing, evaluation, and validation are usually planned during the conceptual design stage and take place parallel to the definition of the overall system design requirements. The testing and evaluation endeavor consists of the testing of discrete components, of various system elements, and then of the complete system as an integrated unit. The idea is to embrace a gradual and ongoing approach that will enable continuous application and enhancement as system design and development progress. Testing and evaluation activities are associated primarily with the design activities and extend through production and construction and then to the system use and support stages. Validation, however, refers to the process needed to ensure that the system configuration as designed meets all specifications. A complete, integrated method should be established for the validation of the system and its elements as an integrated unit. System validation is complete when the system functions effectively and efficiently within its accompanying higher-level system-of-systems composition, hence meeting operational requirements [4].

Central to evaluation is the establishment of comprehensive TPMs, which are predicted and/or estimated values of attributes or characteristics inherent in the design. TPMs are assessed routinely through the stages of the systems engineering process. TPMs may comprise such quantitative factors as mean time between failures, utilization rate, availability, human factors, size, and weight. TPMs arise primarily from the establishment of system operational requirements and the maintenance and support model. They may come from various sources, as with design criteria, and they cover a breadth of issues, but sufficient attention should be devoted to creating human performance TPMs that evaluate the integration of human error reduction tools with the technology necessary to achieve the functionality related to the system's purpose. These human performance TPMs should be created by and will most likely be assessed by plant operations personnel [4, 7].

Human performance TPMs are derived from error precursors,[2] also referred to as performance-shaping factors. Error precursors are unfavorable conditions that generate divergences between a task and the individual. Error precursors hinder successful performance and increase the likelihood for error. By definition, they exist before an error occurs. Error precursors are unique to the situation, the plant, and the personnel involved, and therefore, they require systematic evaluation, logical selection or generation, and potential modification before their adaptation into human performance TPMs [4, 7].[3]

TABLE 6.1 Sample Prioritization of Technical Performance Measures [Reproduced with Permission of American Society of Mechanical Engineers, from "Proactive Human Error Reduction Using the Systems Engineering Process," ASME Journal of Nuclear Engineering and Radiation Science, Corrado, Jonathan, 8(2), 2022; Permission Conveyed through Copyright Clearance Center, Inc.]

TECHNICAL PERFORMANCE MEASURES	*QUANTITATIVE REQUIREMENT (METRIC)*	*RELATIVE IMPORTANCE (USER DESIRES) (PERCENT)*
Absence of Confusing Displays and controls	0 Significance Level 1 human errors	7
	Less than 1 percent Significance Level 2 human error rate per year	
	Less than 5 percent Significance Level 3 human error rate per year	
Absence of hidden system response	0 Significance Level 1 human errors	8
	Less than 1 percent Significance Level 2 human error rate per year	
	Less than 5 percent Significance Level 3 human error rate per year	
Alternative indications	0 Significance Level 1 human errors	11
	Less than 2 percent Significance Level 2 human error rate per year	
	Less than 8 percent Significance Level 3 human error rate per year	

TABLE 6.1 *Continued*

TECHNICAL PERFORMANCE MEASURES	*QUANTITATIVE REQUIREMENT (METRIC)*	*RELATIVE IMPORTANCE (USER DESIRES) (PERCENT)*
Operator proficiency and experience	0 Significance Level 1 human errors Less than 1 percent Significance Level 2 human error rate per year Less than 5 percent Significance Level 3 human error rate per year	14
Operator knowledge of system	0 Significance Level 1 human errors Less than 1 percent Significance Level 2 human error rate per year Less than 5 percent Significance Level 3 human error rate per year	17
Absence of work-arounds/decreased propensity for out-of-specification instruments	0 Significance Level 1 human errors Less than 3 percent Significance Level 2 human error rate per year Less than 10 percent Significance Level 3 human error rate per year	5
Familiarity with task/first time	0 Significance Level 1 human errors Less than 1 percent Significance Level 2 human error rate per year Less than 5 percent Significance Level 3 human error rate per year	19

(Continued)

TABLE 6.1 *Continued*

TECHNICAL PERFORMANCE MEASURES	*QUANTITATIVE REQUIREMENT (METRIC)*	*RELATIVE IMPORTANCE (USER DESIRES) (PERCENT)*
Absence of repetitive actions/monotonous operation	0 Significance Level 1 human errors Less than 2 percent Significance Level 2 human error rate per year Less than 7 percent Significance Level 3 human error rate per year	9
Scarcity of irrecoverable acts	0 Significance Level 1 human errors Less than 1 percent Significance Level 2 human error rate per year Less than 3 percent Significance Level 3 human error rate per year	10
Total		100

NOTES

1 For more on this incident, see David D. Woods, Sidney Dekker, Richard Cook, Leila Johannesen, and Nadine Sarter, *Behind Human Error* (Farnham: Ashgate, 2010).

2 For a complete listing of error precursors, see DOE-HDBK-1028-2009, *DOE Standard Human Performance Improvement Handbook, Volume 1: Concepts and Principles* (U.S. Department of Energy, 2009), pages 2–35 to 2–37.

3 By way of illustration, Table 6.1 presents hypothetical results from a human performance TPM identification and prioritization effort involving a team of individuals representing the designers, operations personnel, and key management

personnel. The quantitative requirement of human error significance levels refers to a defined severity level grouping for potential issues and is discussed in greater detail in Chapter 7 [3].

In this example, the performance factors of Familiarity with Task/First Time, Operator Knowledge of System, and Operator Proficiency and Experience are the most critical, so emphasis in the design process must be directed to these items with respect to human performance [3].

REFERENCES

1. Republished with permission of American Society of Mechanical Engineers, from Corrado, Jonathan. 2021. "The Intersection of Advancing Technology and Human Performance." *ASME Journal of Nuclear Engineering and Radiation Science* 7(1). doi:10.1115/1.4047717; permission conveyed through Copyright Clearance Center, Inc.
2. Proctor, Robert W., and Trisha Van Zandt. 2008. *Human Factors in Simple and Complex Systems.* 2nd ed. Boca Raton, FL: CRC Press/Taylor & Francis Group.
3. Perrow, Charles. 1999. *Normal Accidents: Living with High-Risk Technologies.* Princeton, NJ: Princeton University Press.
4. Republished with permission of American Society of Mechanical Engineers, from Corrado, Jonathan. 2022. "Proactive Human Error Reduction Using the Systems Engineering Process." *ASME Journal of Nuclear Engineering and Radiation Science* 8(2). doi:10.1115/1.4051362; permission conveyed through Copyright Clearance Center, Inc.
5. Sehgal, Bal Raj, ed. 2012. *Nuclear Safety in Light Water Reactors: Severe Accident Phenomenology.* Waltham, MA: Elsevier/Academic Press.
6. Blanchard, Benjamin, and Wolter Fabrycky. 2011. *Systems Engineering and Analysis.* 5th ed. Upper Saddle River, NJ: Pearson Education.
7. Republished with permission of American Society of Mechanical Engineers, from Corrado, Jonathan. 2022. "The Incorporation of Human Performance Improvement into Systems Design." *ASME Journal of Nuclear Engineering and Radiation Science* 8(2). doi:10.1115/1.4051792; permission conveyed through Copyright Clearance Center, Inc.
8. Carlson, Carl. 2012. *Effective FMEAs: Achieving Safe, Reliable, and Economical Products and Processes Using Failure Mode and Effects Analysis.* Hoboken, NJ: John Wiley & Sons.

7

The Incorporation of Procedure and Training Development in the Systems Engineering Process as a Method of Human Performance Improvement in Nuclear Facility Operation

DOI: 10.1201/9781003346265-7

PROCEDURE AND TRAINING DEVELOPMENT IN THE SYSTEMS ENGINEERING PROCESS

The stereotypical perception of human error is that it indicates a flaw present in the human and initiates an undesirable consequence. This misconception places the obligation to prevent such consequences solely on the human. Industry leaders guided by this erroneous understanding continually try to remediate humans' incorrect actions within the system. This leaves organizations and their employees struggling to achieve perfect task performance and always needing to be "more careful." Furthermore, formal corrective actions in response to the error take the form of increased training, reinforcing management's expectations, and occasionally punishment. If these methods are applied to a qualified and experienced employee, they produce peripheral performance enhancements at best, narrowly focused and rarely long-lasting.

As discussed in Chapter 3, human error and error rates are a reflection of mental response to a task, and I noted Rasmussen's three modes of task accomplishment based on the mental processing behaviors exercised at each level: skill-based, rule-based, and knowledge-based [1]. Skill-based tasks are made up of very familiar actions performed in a well-known environment. The human being is virtually on autopilot. Error rates are approximately 1:1,000. Rule-based tasks are known to the operator. Upon accurate recognition of a situation or condition, the performer can apply a known rule to navigate toward a known end objective. Tasks in this performance mode are inclined to follow "if-then" logic, and error rates are approximately 1:100. Finally, knowledge-based tasks are new, unfamiliar, or unique to the performer. Successful performance of a knowledge-based task depends heavily upon the performer's fundamental knowledge, diagnosis, and analysis skills. Unlike the case of rule-based tasks, the operator is not able to navigate toward a previously known end objective. These tasks are best defined as trial and error. Error rates are generally 1:2 [2].

The performer's comprehension of the task, not just the task itself, determines how and at what rate errors are made. An activity could be rule-based for one operator but knowledge-based for another. Therefore, considering the human-machine interface solves only half of the equation; the human-task interface must also be taken into consideration. Substantial gains in Human Performance Improvement (HPI), and ultimately the bottom line, will be achieved only when we match machine to operator in an atmosphere in which the operator can thrive. To fully understand this, it is important to distinguish between errors and events.

Does an organization that seeks improvements in safety initiate a program of error reduction or event prevention? Preventing human error necessitates strict control over external and internal human factors. This control is outside the reach of organizations and can be attempted only through research and precisely controlled examination. The organization that adheres to a safety-conscious work environment concept should make every effort to understand the factors affecting human error rates and associated liabilities, and it should strive to minimize, to the greatest extent possible, human error.

Events are immediately apparent from an organizational perspective. Consequences of these events will steer organizational priorities and provide necessary resources. From this outlook, error can be perceived as a symptom of an event made possible by procedures, processes, and training that are not suitable to protect against human imperfection. Processes, procedures, and training can be easily analyzed and dissected, whereas the human mind tends to be perplexing from an error prevention standpoint. Moreover, events tend to be repeated given similar circumstances and known causes, whereas identical errors rarely yield similar results.

So, the question now becomes: How do you minimize errors? Rasmussen examined human error and recognized that two aspects must be taken into consideration in this regard: human-machine and human-task. The classical view of human error would contend that any faults in the machine or task are also present due to human error. This is unquestionably true. In business, however, that perspective is only valid to the extent that the business has influence over the task and machine. The business organization can use the same method in event prevention with respect to these components irrespective of its influence, bolstering the benefit of undertaking event prevention tactics.

After a new system has been designed and installed, whether due to economic considerations or necessity, the plant at that point has essentially inherited the flaws intrinsic to that system. Although conceivable latent human error embedded in the design or construction of the component can and should be addressed where appropriate, there are limited cost-effective avenues to proactively predict, evaluate, and address such deficiencies once they have been installed and prior to emergence. Transitioning from system design and development to system operation launches the use of operations procedures and reliance on operator training, thereby shifting the focus from the physical system to the operation of that system. This stage of the systems engineering process is the longest-lived and is the primary recognized motivation for the process (after all, the system is designed to be operated). For this reason, operations procedures and training should not be an afterthought to system design, but an integral component of it. Examination of the human-task relationship in system design should begin at the conceptual design stage in the systems engineering process and evolve as the process progresses [3, 4].

Careful Procedure Writing

Beginning with the task, the plant needs to devote considerable attention and due diligence to the generation of processes, programs, and procedures associated with system operation. Specifically, the preparation of operation procedures, a written sequence of steps that establish, maintain, or restore the plant within acceptable operating limits, should begin at the conceptual design stage in the systems engineering process, and the procedures should be further developed and refined as the process advances. This will ensure that procedures are properly developed and that when system production is complete, management is in a position to promptly commence operations. Proper application of human error prevention tools and techniques needs to be soundly intertwined into the framework of these documents and programs. Once generated, these programs, procedures, and processes must be systematically verified and validated to ensure not only adequate system operation, but also that operation remains free of human error trips and traps through appropriate consideration of the task portion of the human-task relationship [3].

Knowledge-Based Operator Training and Drill Program

There is a limit to the amount of detail and information that can be built into procedures. If procedures were written for a person to operate a complex piece of equipment for which they had no prior training or background, they would be much too long, detailed, and convoluted for a reader to follow them sufficiently [3].

For procedures, processes, and programs to be effective, the operator must possess a minimum level of prior knowledge. This necessity leads to an examination of the "human" portion of the human-task relationship. Applying a systems engineering approach to the development of a suitable mechanism enabling humans to adequately and safely operate new technologies is a very complicated and confounding task. When the focus is on the operator, the foundation of safe operation now becomes effective and thorough operator training [3].

Operator training can be divided into two general categories: skill-based and knowledge-based. Skill-based learning concentrates on developing and applying specific skills and behaviors. Learners spend most of their training time learning, developing, and practicing skills through a variety of hands-on, real-life scenarios. Skill-based training will fall short if insufficient time is dedicated to application of the skills and behaviors during the training. The ultimate objective of skills training is to enable the learner not just to acquire proficiency in the skill, but to have the confidence to apply it competently on the job. However, knowledge-based learning is designed to enable the learner to move facts, information, process understanding, and other knowledge from short-term to long-term memory. Much knowledge-based training falls short of this goal due to poor engagement. If the learner is focused more on just finishing the training than on actively trying to assimilate it for further use, the learning impact will be minimal [3].

When learning to manipulate technologically advanced, complex systems, operators should be trained using the knowledge-based approach to ensure adherence to design boundaries, efficiency in operation, and an adequate safety margin. This training should not only communicate the particulars of a system, its components, operation, procedures, and processes, but it should also teach deeper, underlying fundamentals and theory. This ensures a thorough, deeply rooted understanding of not only the specific task and components manipulated, but the impact of that task and manipulation on the system as a whole. Armed with this deeper understanding, the operator can anticipate expected plant response and quickly detect and react to abnormal situations that could turn into plant events. Again, the objective of the operator training program should be to drive performance into the skill-based mode. As with the generation of operations procedures, generation of operator training programs should begin at conceptual design and be further developed as the systems engineering process advances [3].

A vigorous plant casualty control drill program should also be included. A well-developed and consistently administered drill program can effectively provide training and evaluation of facility operating personnel in controlling abnormal and emergency operating situations involving the newly designed system. To ensure that the drills are fulfilling their intended purpose, there should be evaluation criteria for assessing operators' knowledge and skills. Training and evaluation of staff skills and knowledge such as component and system interrelationships, reasoning and judgment, team interactions, and communications can be accomplished through drills.

Proper response to abnormal conditions is vital to ensure personnel safety and protect facility equipment and the environment. Personnel must be able to take the immediate actions necessary to safely mitigate the consequences of an unexpected or abnormal and potentially dangerous condition involving the newly designed system. Drills focus on the actions necessary to respond to these abnormal conditions.

The primary objective of a drill program is to train and qualify personnel. To successfully achieve this goal, drill participation should be integrated into initial and continuing training. An effective drill program is one of the best means available to ensure that the operating staff can safely deal with unplanned and potentially hazardous situations. As with the development of procedures and training, the drill program involving the system in design should be developed during the conceptual design stage and revised iteratively as the systems engineering process progresses. In addition to training operations staff in casualty control with respect to the new system, drill scenario development provides another avenue for system design review and evaluation by having additional sets of eyes view the process from an alternative perspective. This early deliberation of system casualties and necessary operator response not only guides the development of the drill program, but also provides necessary feedback to system designers on desirable improvements and cultivates the continued development of operator training and procedures. Again, the earlier this activity is initiated in the systems engineering process, the better the results will be for system design and operator training.

The rigor and detail of a drill program will differ with facility intricacy and hazard potential. For example, a drill conducted at a reactor facility may involve several people and require a high level of detail, whereas a drill at a site support facility may comprise only a few people and necessitate less detail. Drills on safety-related systems or components at high-hazard facilities may require a large drill team using a detailed drill scenario; drills conducted on safety systems at a low-hazard facility may require a drill team of only a few persons.

To ensure proper implementation of a drill program, the duties, roles, and responsibilities of personnel involved and the mechanics for conducting the drill should be delineated. This ensures consistency of development, conduct, evaluations, critiques, and feedback into the training and drill programs. Alternative methods of conducting drills should be included as an integral part of the drill program to ensure that it is fulfilling its intended mission of training facility operating personnel. Facility management should determine the appropriate level of effort and resources to implement each element of the drill program, consistent with the risk and complexity of the facility.

Regardless of the size, complexity, and risks of a facility, an effective drill program should include the following essential elements: developed drill scenarios, trained drill team personnel, protocol for drill conduct, criteria for drill evaluation, drill critiques, incorporated feedback from drills, and alternative methods of conducting drills.

OPERATOR ATTRIBUTE DETERMINATION

The determination and development of necessary operator skills and training requirements should begin at the conceptual design stage in the systems engineering process and be refined as the process advances. These operator attributes can even manifest as system requirements determined during requirements analysis. At the conclusion of each stage, in addition to adjusting operations procedures and training, employee critical skills, knowledge, and education requirements should be examined and, if necessary, improved. Workforce planning should be built into all stages of the systems engineering process to ensure that personnel with the necessary attributes are available when needed. Human development should be considered as important as system design. Workforce design is too often considered late in the systems engineering process and the initial plant startup and early operations suffer due to inadequate workforce planning or excessive costs associated with accelerated operator training and qualification.

Embedded in the determination and development of the required workforce are population studies to canvass the local population for demographics such as industry expertise, education levels, and education ability. With this baseline assessment of the population, training programs can be efficiently developed in view of the general strengths and weaknesses of the available workforce, and specific needed skill sets can be assessed, targeted, and fostered if necessary. Nonetheless, as discussed above, this needs to happen early and iteratively so that issues can be addressed and corrected during design.

Another important subtlety in workforce development is the development of leaders. As the systems engineering process proceeds and, more generally, as the speed of business increases and the complexity of technical advances rises, it is easy to focus only on immediate needs and pay less attention to the systemic issues that ultimately drive long-term success. As the process progresses, management needs to be constantly assessing the connection between leadership practices, employee work passion, customer devotion, and the bottom line. There is a clear connection between the quality of an organization's leadership practices, as perceived by employees, and employees' intentions to stay with an organization, perform at a high level, and apply discretionary effort.

SYSTEMS ENGINEERING INFRASTRUCTURE

The design and development of an innovative system necessitates an adaptive and equally unique and innovative systems engineering infrastructure. One should not fall into the trap of developing a twenty-first-century product in the framework of twentieth-century systems and processes. It is important to remain apprised of and consistent with industry standards and best practices with respect to organizational design, management processes, software development and functionality, and administrative methods—not necessarily with the techniques that everyone has used in the past and is comfortable with. Business is always evolving and striving to reach new heights; the infrastructure should stay in lock step with this evolution to ensure the development of the best product or system using the best available resources.

As the systems engineering process for product or system development is launched, the proper infrastructure should be established to support it efficiently. The infrastructure design should be a formal, guided process of integrating the people, information, and technology of an organization. It is used to match the form of the organization as closely as possible to the purpose that the organization seeks to achieve. Through this design process, organizations act to improve the probability that the collective efforts of their members will be successful.

THE FINAL PRODUCT

Operator involvement in the systems engineering process; human performance integration with system operational requirements and system testing, evaluation, and validation; the implementation of a strong knowledge-based operator-training program, including a drill program; the generation of well-written and validated procedures; and the selection and cultivation of personnel aptly prepared for system operation—all these actions are connected with a common theme: the development of a well-qualified, capable, and equipped human operator for the given system. As a clear illustration, we can recall the swift, logical, reasoning used by an officer in the Soviet Air Defense System to save the United States from nuclear war.

On September 26, 1983, Lieutenant Colonel Stanislav Petrov of the Soviet Air Defense Forces was on duty at the command center of the Soviet early warning satellites. Petrov's duties included monitoring the satellite early warning system and informing his chain of command of any nuclear attack. If the early warning system indicated an attack, the Soviet Union's response would be an immediate counterattack. Close to midnight, the early warning system indicated five inbound intercontinental ballistic missiles from the United States. Petrov deemed the detection to be an error, since a first-strike nuclear attack by the United States, should one be mounted, was expected to involve hundreds of concurrent missile launches with the intention to incapacitate any counterattack. In addition, the early warning system had been newly designed and installed. In his estimation, the system was not yet entirely reliable. Petrov also observed that the ground radar had failed to corroborate the indications of the early warning system. Petrov ultimately rejected the warning and classified it as a false alarm. It was later determined that the false alarm had been produced by an atypical orientation of sunlight on high-altitude clouds and the early warning systems satellites' orbits [5].

Imagine if the Soviet Air Defense System had been designed with little to no human interaction and had automatically initiated a counterattack based on the indication received by the system and the Soviet Union's established response strategy. This dramatic example emphasizes the importance of the human being within a system who is well trained and knowledgeable of the system that they operate.

WAYS TO MINIMIZE THE IMPACT OF HUMAN ERROR THROUGHOUT THE SYSTEMS ENGINEERING PROCESS

As I have emphasized repeatedly, human error cannot be entirely prevented, but if proper tools and techniques are implemented early in the systems engineering process, it can be reasonably minimized. With this goal in mind, it is also important to establish a system to minimize the impact or potential impact of errors. Along with the steps discussed previously in this chapter, an HPI process should be established not only to remain cognizant of human performance during system design, but also to provide a means to evaluate human performance in operating the system at its various stages. This process incorporates HPI logic into system design and subsequent system operation and provides a mechanism to assess and mitigate potential impacts of human error during system design. This process should be established early in the systems engineering process to identify how human error affects plant systems and equipment; this information in turn will provide necessary feedback into system design process [4]. This process is displayed by means of a use case model in Figure 7.1. The nucleus of this model is converging the efforts of all project entities on meeting the established design requirements comprised of safety, security, and customer requirements.

To institute this process in a manageable fashion, organizations should determine human error severity criteria and establish a tracking system to capture these potential human error-induced issues for consideration in developing engineered and administrative component features, human error prevention training and reinforcement, and lessons learned. These potential issues and incidents can be discovered at any stage of the systems engineering process, during the development of procedures for system operation, during

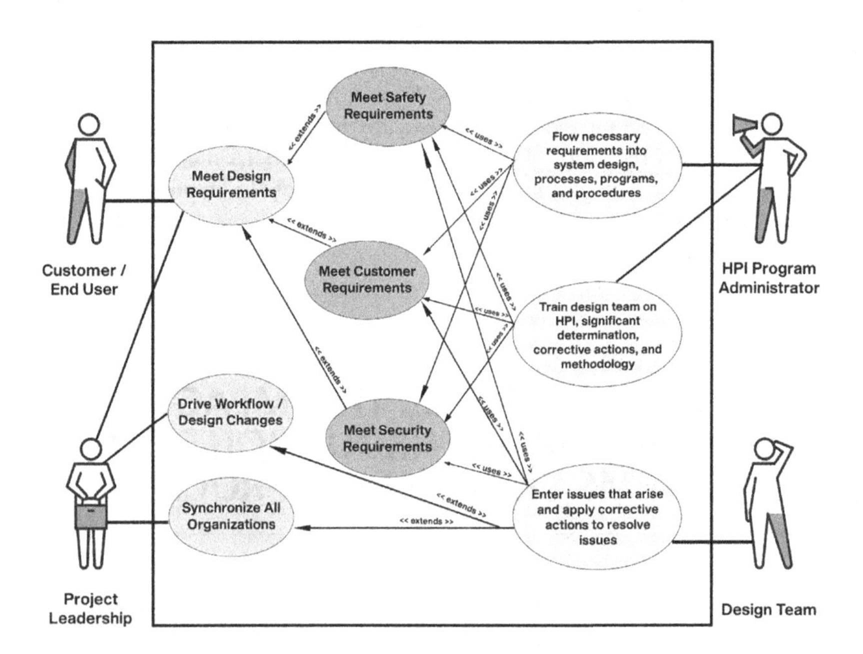

FIGURE 7.1 Use Case Model of the Human Performance Improvement Process [Modified from Corrado, Jonathan. 2017. "Technological Advances, Human Performance, and the Operation of Nuclear Facilities," PhD Dissertation, Colorado State University, Fort Collins, CO, ProQuest (AAT 10258407)].

training development, during Failure Modes and Effects Analysis, and during the design reviews conducted at the conclusion of each stage. In addition, this information can be used to determine trend data and thereby predict negative behaviors before potential errors or adverse events can occur. These data can also provide a multitude of information to be used for other purposes across the company, industry, and systems engineering discipline. For example, they can be used to track and document trends in potential human errors during system design and actual human errors during system use to provide feedback into the operator training program (for either continuing training or the qualification process), to generate reports for project and organizational management, to determine organizational goals, or to compare plant performance to that of other plants in a similar industry [4].

First, organizations should create a formal process to examine incidents and perceived vulnerabilities that occur throughout system development during the systems engineering process and, later, during plant operation. A methodical approach should be established to systematically examine the potential incident or vulnerability through a series of questions so as to determine regulatory impact and safety significance, determine which organization will perform the corrective action or causal analysis, and ultimately correct the deficiency. Figure 7.2 displays an example of a process chart exhibiting the progression from problem reporting to problem correction [4].

As displayed at the top of Figure 7.2, a determination of the significance level should be made. I offer an example of the classification of severity criteria tailored for a nuclear facility in Table 7.1, where Level 1 events are the most severe and Level 3 are the least severe.

With the human error severity criteria determined, a tracking database established, and a formal process to collect and analyze the information, organizations now need to determine acceptable quantities of error within the established levels, so that this information can be fed into TPM quantitative measurement determination and assessment. The goal of this system should be to keep potential errors at the lowest significance level. So how is this done? The answer is straightforward but requires consistent attention. Due diligence, constant emphasis, and management support must be devoted to the identification and reporting of potential human errors and the corrective action processing of the problems as they are reported (per Figure 7.2). This processing will determine the significance level of the error and therefore the required amount of staff attention. Deployment of sufficient corrective actions will not only fix the immediate problem but will also prevent similar potential errors from occurring later in the systems engineering process. Depending on the significance level, a generic implications review could be conducted to determine the extent of the condition, the extent to which the actual condition or similar conditions exist in other plant processes, equipment, or human performance,

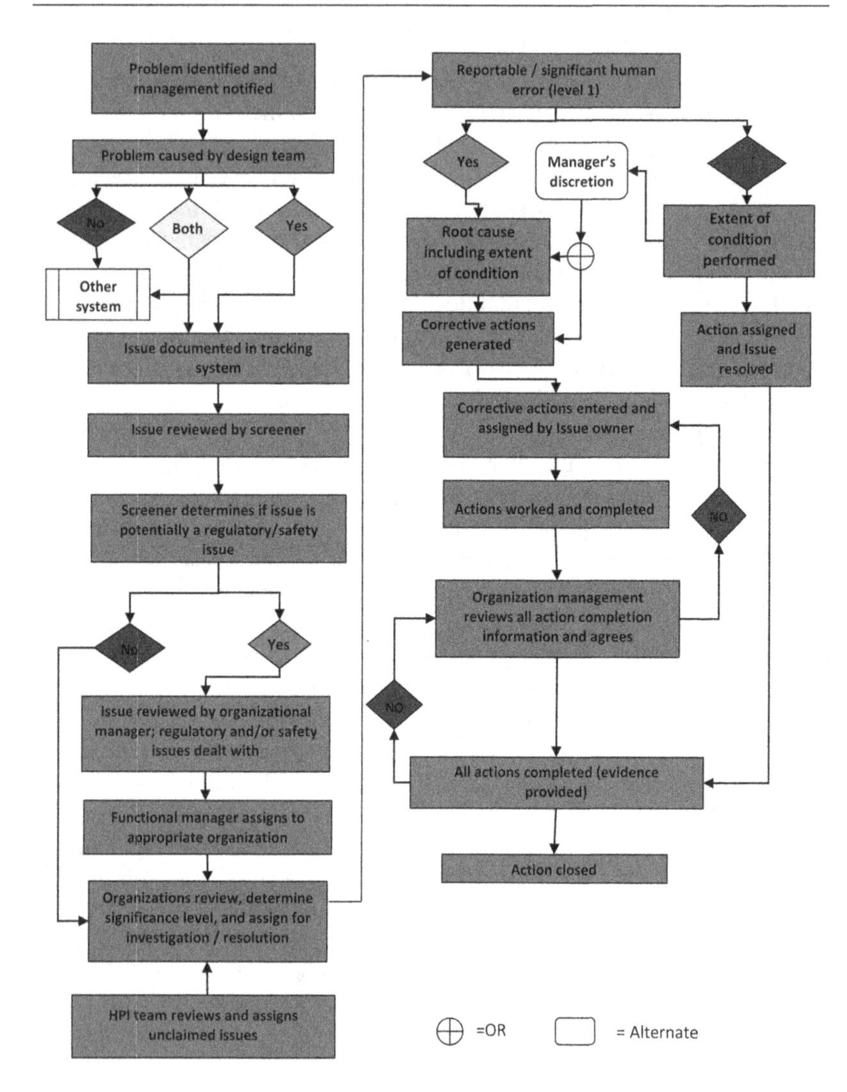

FIGURE 7.2 Human Performance Flowchart [Modified from Corrado, Jonathan. 2017. "Technological Advances, Human Performance, and the Operation of Nuclear Facilities," PhD dissertation, Colorado State University, Fort Collins, CO, ProQuest (AAT 10258407)].

TABLE 7.1 Severity Criteria Matrix Example [Modified from Corrado, Jonathan. 2017. "Technological Advances, Human Performance, and the Operation of Nuclear Facilities," PhD Dissertation, Colorado State University, Fort Collins, CO, ProQuest (AAT 10258407)]

INCIDENT CATEGORY	*LEVEL 1*	*LEVEL 2*	*LEVEL 3*
Plant transients	Repeat occurrences of organizational or programmatic breakdowns that affect nuclear safety. (Level 1, 2, or 3) Significant event requiring use of safety features Multiple equipment malfunctions or human errors occurred that significantly increased the severity of the transient. Significant operating/design violations of safety analysis	Failures that could affect multiple safety systems or components Misvalving operation or maintenance error on wrong equipment causing tripping or transient on operating equipment NRC reportable event requiring a written response Significant program weakness in design, analysis, operation, maintenance, testing, procedures, or training.	Minor program weaknesses in design, analysis, operation, maintenance, testing, procedures, or training identified by independent or management assessors An auxiliary plant transient M&TE that fails calibration but does not cause operational impact to safety-related equipment Safeguards/security issues that do not meet the regulatory criteria.
Personnel safety	Death not due to natural causes Major disability injury	Injury or near miss with fatality potential Work-related injury requiring inpatient hospitalization Individual exceeds regulatory dose limits Multiple or other substantial personnel contamination instances	Personnel contamination events occurring from procedural violations or poor radiation worker practice. Personnel contaminations due to human error and which result in dose assignment Defeated or missing LOTO with no potential for exposure to hazardous energy

(*Continued*)

TABLE 7.1 *Continued*

INCIDENT CATEGORY	*LEVEL 1*	*LEVEL 2*	*LEVEL 3*
Environmental impact	Releases resulting in significant threat to human health or environment EPA violation or OSHA citation that results in enforcement action Release in excess of radiological limits with actual or potential for off-site impact	Immediately reportable spills with potential to harm environment Release in excess of radiological limits with no potential for off-site impact Repeated failures such as spills of chemicals or oil, improper storage, failed secondary containments.	Permit criteria threatened by a discharge Failure to maintain secondary containment for chemical/oil spills Isolated non-reportable failures such as spills of chemicals or oil, improper storage, and failed secondary containments.
Economic/ operational impact	Missing business commitment	Extensive equipment damage (e.g., required replacement or substantial repairs) Conditions resulting in substantial outage delays or extensions Lengthy unplanned outage or operation at substantially reduced production Repeated failures to implement or maintain commitments to regulatory agencies	Fire protection equipment unavailable when it was needed Conditions detected by independent or management assessors that do not represent a substantial organizational or programmatic barrier breakdown Adverse trends in equipment, programmatic, or human performance that do not directly challenge safety, regulatory compliance, or reliability

and the extent to which the root cause and contributing causes of an identified problem could impact the same or similar plant processes, equipment, or human performance (both irrespective of severity level). Based on this review, the organization could put additional preventative actions in place to inhibit error-likely situations. This process, if effectively performed, will constantly reinforce good human performance practice, reduce human error in system design and potential human errors during operation, proportionally decrease the volume of potential problems at all levels, and continually drive errors to the least significant level [4]. This concept is displayed visually in Figure 7.3.

In conjunction with a sound HPI program, a management team should be established to ensure that the program is effectively and consistently implemented with a particular emphasis on evaluating significant issues, adverse trends identified, and ineffective corrective actions that were applied to conditions adverse to safety and security. This team could also incorporate all issues, not just human errors, into its engagement. Specifically, management oversight is required to ensure the following:

- Proper disposition of issues
- Application of the proper significance level
- Proper trend identification

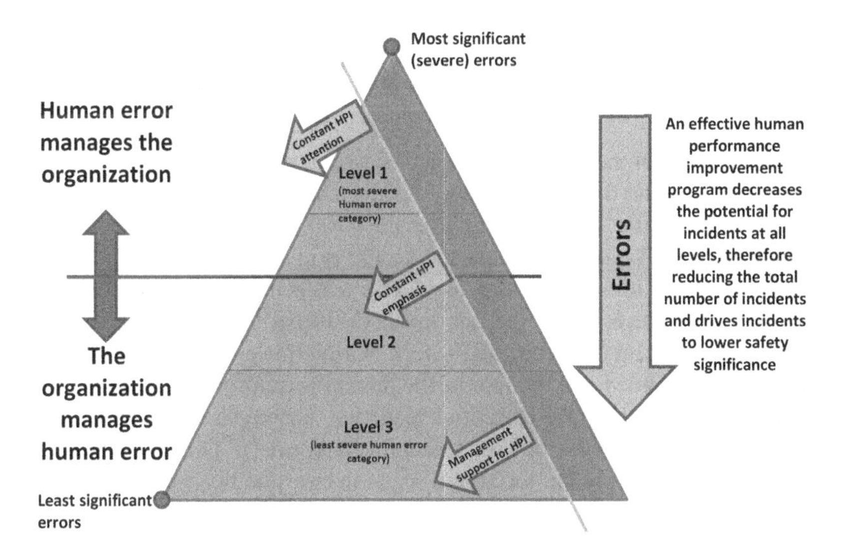

FIGURE 7.3 Human Performance Significance Level Triangle [Modified from Corrado, Jonathan. 2017. "Technological Advances, Human Performance, and the Operation of Nuclear Facilities," PhD Dissertation, Colorado State University, Fort Collins, CO, ProQuest (AAT 10258407)].

- Greater management intervention in certain issues
- Identification of issues that should be disseminated as lessons learned
- Proper application of corrective actions
- Periodic review of performance indicators
- Differentiation between corrective actions and process improvements
- Review of corrective action plans to ensure that they address previous ineffective corrective actions
- Review of root cause investigation results

The reporting of human errors should aim to produce a culture focused on safety and operational excellence. Employees should be not only encouraged but expected to make error incident reports. This should be a living part of the organizational DNA and "business as usual" for plant employees. Most important, employees must know that they will not experience retaliation for making reports. The generation of a positive nuclear safety culture and a safety-conscious work environment is a delicate and complex task, but once it has been established, it will improve morale, breed content employees, strengthen plant safety, and ultimately boost productivity.

DESIGN OF NEW SYSTEMS

While the focus of the previous section discusses human error reduction while improving or upgrading existing systems, technologically advanced systems and facilities of the future nuclear fleet need to keep a steady press to keep human error as low as possible. As the evolution of systems progress, the role of the humans within the system inevitably will continue to evolve as well. This evolution necessitates the continued drumbeat of HPI while keeping sights set on system efficiency and safety. Within this context, as new future systems are designed and the role of the human is transformed, human error should be architected for substantial reduction as compared to predecessor systems with an increase in system stability and safety. This necessitates a different focus appropriate for a new set of nuclear facility technology with new automation types and the roles of humans modified accordingly to accommodate. This focus should compel the optimization of human-machine roles to reduce human error and increase inherent stability of the system [7].

As new systems are designed, systems engineers must view the system with an open mind and must give consideration to various technology options, affordability, accept the fact that human interaction is inevitable, account for human limitations, and, ultimately, establish appropriate requirements for the system and for the human. The incorporation of these elements will appropriately balance the next-generation system and lead to an optimized human-machine system [7].

Human-Machine System Optimization

Human-machine interaction is the boundary where interactions between humans and machines transpire. The objective of this interaction is to facilitate safe and effective operation and efficient control of the machine from the human end as the machine simultaneously feeds back information that aids the operators' decision-making processes. Generally, the intent of human-machine interface design is to produce a user interface that makes it straightforward, efficient, and manageable to operate a machine so as to attain a desired outcome. This generally means that the operator needs to provide a specific input to achieve a desired output, and also that the machine minimizes undesired outputs to the human. To achieve safe and reliable operation, this interface requires a practical balance between the human and the machine. The purpose of this section is to propose a methodology that optimizes the balance between the human and machine sides of the human-machine interface of a system at a nuclear facility [7].

Background of Human-Machine System Interaction

The very detailed interaction between machines and humans in nuclear facilities is a changing technical matrix that is addressed by means of a human-machine interface. Studies of the interactions between machines and humans have been conducted for more than half a century. Effective models of human-machine interaction and matrices for numerous products can be found in the nuclear industry [8].

The interactions between humans and machines at nuclear facilities have been recognized as significant due to their safety implications; as a result, these interactions are handled with a high level of care. The entire system of the human-mechanical interface, the machine and the human user are described as the human mechanical system (HMS) and may include distinct user categories, such as managers, maintenance personnel, and operators, all of whom have distinct needs regarding information and control.

The term "mechanical" in HMS refers to any category of changing technical system, which can include the software and automation components as well as the mechanical components. The technical system's automation components are described as administrative and control systems. These systems interact in a direct manner with the technical machine interface. The power generation process at a nuclear power facility is an appropriate example.

In a nuclear facility, the decision support systems are state-of-the-art, and the component machines are programmed with substantial knowledge and can provide advice to the user. The application domains of the HMS include human users, the human-machine interface, and the mechanical components [8].

The context of automation in the administration of changing technical matrices has been significantly enhanced over the last several years. This statement is true for all the technical systems associated with nuclear facilities and products. Elevated levels of efficiency have been attained by means of the enhanced application of automatic administration. The requirement for communication between machines and humans increased with the augmented context of automation. Usually, enhanced automation does not replace the human's interaction with the machinery; rather, it transforms the location of the interface between the two [9].

The Human-Machine Interface

The bread and butter of nuclear facility operation is safety and efficiency. In the fulfillment of these essential elements, there is a recurrent concern for efficient human interaction with the machines being operated. Enhancement of this human-machine link is a necessary enterprise within the process management not only for the organizational overview, but also for the schematic placement of roles and specific defining variables for maintaining the interaction. Considerable effort should be applied to reach a level of optimization where this human-machine synergy takes into account the human-machine interaction as part of routine operations during production, maintenance evolutions, and off-normal operations requiring emergency mitigation, such as the impact of natural phenomena or unexpected failures within the system. To operate effectively, each human-machine link must be well defined and specifically engineered using the requirements established to fulfill the system purpose, which necessitates appropriate design decisions that meet this purpose and design that accounts for specific roles and processes remaining integrated and strong. This implies not only considering all the decisions, manipulations, and potential errors that can be made in the system under design, but also lessons learned from other similar systems [7].

To achieve safety and efficiency in operations, the human-machine interaction should be a focal point during system design. This interaction has many facets and factors to consider. An integrated approach to seeking further process dynamics remains defined by interaction and does not leave processes on their own to facilitate operations independently. Within the human-machine link, the machine is designed and installed to fulfill a specific purpose and, in doing so, must provide a mechanism to exhibit required system parameters so that the human can intervene to ensure that the machine is operating at the necessary capacity for efficient functionality and take necessary actions if an unsafe condition arises. Furthermore, conditional models for both emergency and nominal routines aimed at meeting safety and operating standards see a network of knowledge centered upon creating processes that diminish uncertainty and strive to define every contingency. Systems should be designed with machine monitoring in mind that allows the human activity to be fixed and defined by routine that eliminates much uncertainty in operation [7].

Human Error and the Human-Machine Interface

This study has repeatedly emphasized the considerable potential for human error, and this is especially true within the context of the human-machine interface. Specifically, there is a greater potential for errors in the case of information interpretation that requires both humans and machines to collaborate. The error emerges when erroneous information defines the process. Without the correct, most relevant system information, errors can arise both from the action of internal machine control mechanisms or human analysis and from the resulting actions based on wrong information. It is important to design the system for a stronger human-machine interaction with less censoring of human behavior and more mining of information for evaluation of processes; this allows the human to incorporate change in ways that the machine cannot. One must be wary, however, of too much ingenuity and flexible decisions as a means of deviating from the original purpose of operation. To redefine safety for the wrong needs also suggests that the wrong values are applied, which can result in human error; the resulting consequences, in the context of nuclear facility operation, can be detrimental [7].

Some errors happen because humans interact at a cognitive, rational level but also have emotion and conflict to consider when making choices. In some contexts, errors take place out of complacency or boredom regarding one's role in the system. For evaluation and redesign to take place, not only is it important for each member of the team to be cross-trained in some defined functions, but information also needs to be shared and new knowledge circulated to inspire new ideas for solving core capabilities of safety and efficiency. Roles are aligned with integration because knowledge is open, and this leads to systems that have flexibility and tolerance of the human factor of emotion. Design of the system should not be dependent upon freedom or correct roles but on how the process is defined to carry out the purpose. Roles fit into the process and so do human personalities for specific roles [7].

Decisions and the Human-Machine Interface

Decisions can be challenging for the human even if machines monitoring for system reliability show little change in data or information about the system. Decisions can be made for the wrong reasons based on limited information. The data may not match the task, or the outcomes may have limited validity. Data should reflect what the operator understands about the activity, behaviors, and resources that define the process. Unfortunately, at times incorrect observations, missing information, or misinterpretations of information can result in poor decisions. In addition, operators may evaluate the data from the interface differently from each other, creating a lack of collaboration in optimizing the links in the system. The machine may offer valid knowledge, but its translation into human knowledge is uncertain, potentially reducing the information's value for making decisions. So even if there may be adequate monitoring systems, backup data systems, cameras, institutionalized protocols for data management, and suitable storage and handling, the information can be interpreted incorrectly, leading to poor decision making and reflecting weakness in design [7].

Taking into account the human element will permit use of decision-making systems to consider operational scenarios. Deterministic functions are not flexible and allow for the same variable to be inputted only to return the same outcomes. Non-deterministic functions allow for some flexibility in the sense that they can return different outcomes even with the same input. This allows for further dimensionality in modeling the various operational scenarios within the process to aid in redefining the human-machine interface, based on additional data that support the desired, optimal process outcome. Within the framework of safety, serious consideration must be given to situations where decisions are made quickly, under pressure, and with uncertainty. These operational scenarios exist for all industries, but from the perspective of nuclear operation, where the consequences of errors can be extremely severe, these scenarios should be identified in detail and the human-machine links to be analyzed should be defined [7].

Understanding how system capabilities can define risks to the system related to safety and performance involves a recognition that operators can and must learn from mistakes made and that designers must perceive opportunities from errors. Continued reassessment and an approach to redesigning human-machine links can create system resilience. Errors can be less costly in this context of specific redesign architecture. Learning from past errors can enhance anticipation and expectations with regard to future events in terms of technology control strategy, unleashing its full power usefully to address factors that create conflict or uncertainty. This is an avenue for innovation because it drives the forces of uncertainty away and allows for technology application for human benefit [7].

Considerations in Optimization

Human Interaction Is Inevitable

Although it might seem that the best way to engineer fault-tolerant systems is to eliminate any potential for human error, this assumption could lead to the conclusion that anything that can be automated should be automated. An intrinsic flaw in that argument is that no matter how complex the system, humans still need to interact with it. As discussed previously, when human interaction with a system is reduced, degradation of human skills can result. As a clear example, consider the decline in children's handwriting skills. In the past, all elementary students learned cursive writing. Today, the emphasis has shifted toward computers, keyboarding skills have become more valued, and children are spending less time on handwriting, with a corresponding decrease in skill level. From a systems perspective, we can look at the Air Force as a model. When pilots started flying jets with increasing degrees of automation, researchers found that their overall flying proficiency was increasing, but that they were losing their manual flight abilities.[1] As a result, pilot advisory boards started suggesting that pilots should fly more manual hours in order not to lose those skills [10].

My main focus has been on human error as a result of technological advancement associated with the upgrade of existing systems in existing nuclear facilities. Technological upgrades are unavoidable, so consistent efforts must be devoted toward keeping human error as low as possible amidst those changes. As the evolution of systems progresses, the role of humans within the system will inevitably continue to evolve as well. This evolution necessitates a continued emphasis on HPI as part of maximizing system efficiency and safety. Within this context, as new systems are designed and the role of the human is transformed, human error should be targeted for substantial reduction relative to predecessor systems, with a corresponding increase in system stability and safety. For this purpose, the roles of humans and machines must be optimized in the design and implementation of new types of automation, but it must be understood that human interaction with systems is compulsory and of major importance [11].

The Role of Complex Systems and Human Actors

From a systems engineering perspective, it is important to understand the precise role of complex systems in the interactions between human actors. In the past, computing and cybernetics systems were focused on extending the physical attributes of humans and doing things that humans could not safely do, such as interacting directly with highly radioactive materials. However, as computing power has increased as a natural result of Moore's Law, now systems are advancing into the realm of enhancing cognitive and mental capacities. Reducing human error through system architecture is a very important goal, as was revealed by tragic failures in the use of Patriot missiles in 2003 during the Iraq war. The Patriot radar systems were engineered to record false hits and false alarms without displaying any uncertainty regarding the target. The human tendency when working with automated systems is to trust the accuracy of the information provided. Unfortunately, because the systems were poorly engineered, the humans interacting with such systems took the blame for shooting down a British Tornado and a US Navy F18/A, which were incorrectly identified as targets by the automated systems [10].[2]

As this example illustrates, one crucial part of engineering human error out of systems designs is to make certain that the systems do not introduce their own errors. If humans interacting with engineered solutions are to be expected to operate consistently, the systems must provide sufficient information to the humans involved so that they can use their unique characteristics in analyzing the situation and making the proper decisions. A crucial responsibility of the systems engineer in designing a human-machine interface system, therefore, is to properly articulate who makes the final call in crucial decisions. In some situations, the automated systems should never be overridden by the human operator, such as when doing so would expose the humans involved to unacceptable amounts of radiation exposure or automatic shutdowns due to low or high system pressure. Conversely, in other situations, such as normal or controlled operations, an automated system should not be able to override the human's judgment. A good human-machine interface achieves an appropriate balance between the skills of the operators involved in the particular situation and the inherent strengths of the systems involved [10].

Some areas where human errors can occur in integration with automation systems reflect a lack of knowledge in the area of human cognition and cognitive processes. Humans have amazing but still limited cognitive abilities, and systems designers must take those limitations into consideration. For example, excessive use of multi-windowed systems for monitoring can result in degraded human performance because these systems overtax the operator's attentional capacity.[3] Important alerts could be missed due to exceeding the human's cognitive processing abilities. Whereas an initial alarm can seize an operator's attention, repeated alarms (or so-called nuisance alarms) inevitably cause the operator to become habituated to them (like the classic fable about crying wolf too often). Thus, when a truly important alarm is sounded, the operator may tune it out and not attend to it. Excessive flexibility also presents a problem for human operators. A classic example is the smartphone: most users take advantage of only a small fraction of its functions because most of them are too complex or take too much time to figure out. Therefore, for systems engineers to eliminate human error factors, they must fully understand the cognitive limitations of the users of such systems and compensate for them by designing systems that augment human capabilities of flexible thinking, while at the same time not presenting such an overwhelming variety of options that crucial indicators or tools are ignored because they overwhelm the humans whom they are designed to assist [10].

Account for Human Limitations

As technology advances, it becomes increasingly apparent that from a systems engineering perspective, systems have a (nearly) unlimited capacity whereas humans do not. Therefore, to create systems that reduce human error as much as possible, respect for the limitations of the human brain and its processing capacity is necessary. As I mentioned, the human brain has a limited processing capacity because its processing activity occurs in the area referred to commonly as short-term memory. The short-term memory circuit consists of sensory memory, which can contain a few seconds of data at most, and the short-term memory store. The complexity of designing with respect for this system lies in the fact that the attention capacities of short-term memory are divided between data just taken in, data retrieved from long-term memory for processing, coding procedures, and search strategies.[4] Therefore, when designing complex systems with many things that must be attended to, systems engineers must find a way to narrow the information presented to the most essential elements, so that the crucial information that must have the human operator's attention is front and center at all times [10].

An analogous system in which visual displays play an essential safety role is visual control systems for automobiles. Indeed, the so-called instrument cluster is a crucial part of the safe operation of a motor vehicle because it relays safety-related signals to the driver. As automobile safety systems become increasingly advanced, with more components designed to assist the driver in safe automobile operation (including collision detection, parking assistance, night vision assistance, adaptive cruise control, and more), the space available for displaying this information becomes a key limitation that must be overcome. To deal with the space limitations of automobiles, designers have turned to the novel solution of creating configurable dashboards. One of the salient points arising from this research is that all this additional information must be provided to the driver of the automobile without simultaneously distracting them from the primary goal, which is to safely operate the automobile. This is why human-machine interfaces in cars must be designed in a way such as to prevent human error.[5] Research on automobile interfaces has found that customizable dashboard interfaces increased passive safety, as the ability to tailor the interfaces was correlated with significant improvements in users' attention and reaction capabilities.[6] One reason posited for this improvement in attention is that customized interfaces are closer to the real-world systems that they are supposed to support. Because the end user is directly involved in the customization, there is a reduction of the psychological distance between the user and the system, making it easier for the user to execute the needed tasks [10].

This insight can be directly applied to the arrangement and composition of nuclear facility control interfaces. Vital operational information should be maintained at the forefront (e.g., reactor power, core temperature, pressure), but other information can be managed and displayed at the operator's discretion or according to plant procedures. This arrangement combines the availability of vital information with operator knowledge, based on training and experience, as to what displays are necessary at different times without overwhelming the operator's cognitive capacity [10].

Considerable data suggest that the use of a dashboard concept holds promise in the area of human-machine interface design. One of the most valuable aspects of dashboards is that they can improve decision making (or prevent errors) by amplifying cognition as well as making the most of humans' limited perceptual capacity. Flexibility in selection of dashboard formats aids in user response as well as in accurate use of the information.[7] In addition, dashboards can be an effective answer to the problem of memory overload, a key limiting factor that can contribute to human error in using human-machine interfaces [10].

Several elements of dashboard design must be taken into account when one is designing a dashboard human-machine interface. The first factor is visualization. The information visualized within a dashboard design must actually amplify cognition. The visualization of data can be considered correct if the end users consistently decode the information presented properly. Again, respecting the limited cognitive capacity of short-term or processing memory, an effective dashboard design will strike an appropriate balance between visual complexity and the information utility required for the particular situation [10].

Other functions can be built into dashboards to reduce errors. For example, automated alerts (in limited fashion because of the aforementioned problem of habituation) can be included in the dashboard design, along with theory-guided format selections that can help to lead an operator to the correct selections for a given scenario. Limiting the dashboard to a single page and a simple color scheme, along with links and grid lines for 2D/3D data graphs, is another research-supported way to improve the visual clarity of dashboard designs [10].[8]

Another important aspect of human-machine interface design is the affordance of information in a way that is consistent with the desired results. "Affordance," in the context of human-machine interface design, refers to the features provided to the user. One way in which affordance can reduce human error is in the proper presentation of choices or options. For example, shading a button or menu item in gray and making that choice inaccessible because it is a contextually inappropriate choice can guide the user toward making correct choices in the situation. There is, however, an accompanying danger that affordances can also misinform or misdirect a user into an incorrect choice or option. For example, a horizontal line on a scrolling page could lead the user into thinking that the page has ended when actually there is more content "below the fold." Improper or misunderstood instructions can introduce or increase the opportunity for human error in human-machine interface use [10].

CONFIDENCE IN PROPOSED APPROACHES

To assess experts' confidence in the proposed methods described in Chapters 6 and 7, a Likert-style survey was designed and administered to a group of systems engineers to gain their impressions of the methods proposed, whether the methods would be successful in reducing human error, whether the experts would be likely to use these methods, and issues affecting the implementation of these methods within the systems engineering process in the industry setting. The survey sample consisted of forty system design engineers (age twenty-five to sixty-five) of various backgrounds, locations, and disciplines (e.g., nuclear engineers, electrical engineers, mechanical engineers) who were knowledgeable regarding the systems engineering and design process.

The statements contained in the survey were extracted directly from the sections of this chapter. The respondents were provided with the chapter content and the questions and statements along with the five-point Likert scale. Depending on the question, the potential responses were labeled in one of two ways: (1) "never," "occasionally," "regularly," "frequently," and "always," or (2) "strongly disagree," "disagree," "neither agree nor disagree," "agree," and "strongly agree" [12]. The results are shown in Figure 7.4.

As shown in Figure 7.4, the totals are weighted to "always" or "strongly agree" with no responses indicating "never," "strongly disagree," "occasionally," or "disagree"; therefore, the responses are largely favorable for the methods proposed in this chapter. Additional analysis is shown in Table 7.2.

The percentage agreeing represents those who indicated "strongly agree" or "agree" (or "frequently" or "always," depending on how the question was worded) on the five-point Likert scale in response to each item. The top box represents only those responses of "strongly agree" or "always." The net top box is found by counting the number of respondents who selected the top choice

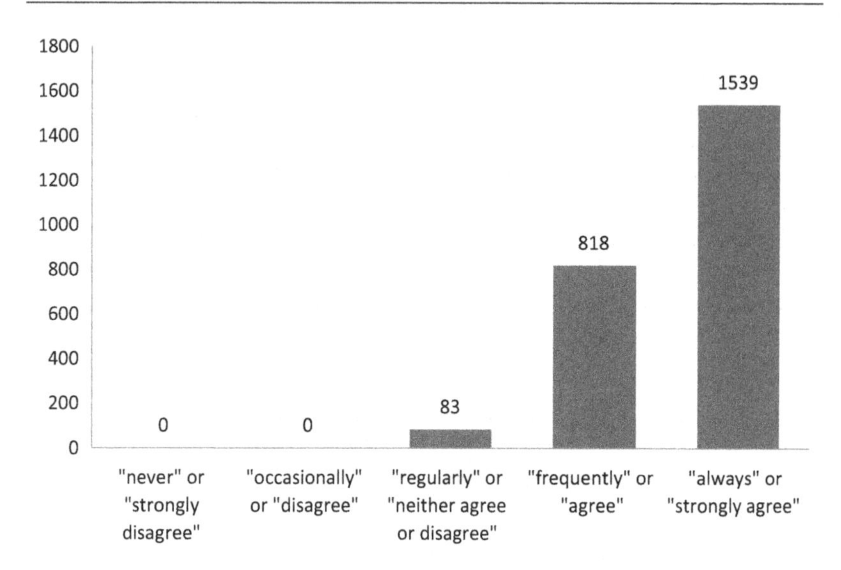

FIGURE 7.4 Likert Quantitative Tabulation [Reproduced with permission of American Society of Mechanical Engineers, from "Survey Use to Validate Engineering Methodology and Enhance System Safety," ASCE-ASME Journal of Risk and Uncertainty in Engineering Systems, Part B: Mechanical Engineering, Corrado, Jonathan, 8(2), 2022; Permission Conveyed through Copyright Clearance Center, Inc.].

TABLE 7.2 Results of Survey Analysis [Modified from Corrado, Jonathan. 2017. "Technological Advances, Human Performance, and the Operation of Nuclear Facilities," PhD Dissertation, Colorado State University, Fort Collins, CO, ProQuest (AAT 10258407)]

ANALYSIS	*RESULTS (%)*
Percent agree	96.6
Top box	63.1
Net top box	63.1
Z-score to %	86
Coefficient of variation	12

and subtracting the number who selected the bottom choice. The *z*-score to percentile rank converts the raw score into a normal score due to the fact that the rating scale means often follow a normal or close to normal distribution. The coefficient of variation is a measure of variability, unlike the first four variations, which are measures of the central tendency [13].

The quantitative results show that the respondents were strongly favorable regarding the value, usability, and likely industry acceptance of the concepts presented in this chapter.

In addition, the respondents were given the opportunity to offer open-ended comments on each section of the survey. All remarks expressed an appreciation of the concepts, an acknowledgment of the benefits of implementing such concepts, and expectations that future implementation would receive a positive industry reception. See the Appendix for a comprehensive explanation of the survey and additional details associated with the results and analysis.

NOTES

1 For more, see Douglas Orellana, and Azad Madni, "Analyzing Human Machine Interaction and Interfaces through Model Based System Engineering Practices," *INCOSE International Symposium* 22, no. 1 (2012): 1780–90, http://doi.org/10.1002/j.2334-5837.2012.tb01436.x.
2 For more, see Azad M. Madni, "Integrating Humans with Software and Systems: Technical Challenges and a Research Agenda," *Systems Engineering* 13, no. 3 (2009): 232–45, http://doi.org/10.1002/sys.20145.
3 See Madni, "Integrating Humans."
4 For more, see John T. E. Richardson, Randall W. Engle, Lynn Hasher, Robert H. Logie, Ellen R. Stoltzfus, and Rose T. Zacks, *Working Memory and Human Cognition* (Oxford & New York: Oxford University Press, 1996).
5 For more, see F. Bellotti, A. De Gloria, A. Poggi, L. Andreone, S. Damiani, and P. Knoll, "Designing Configurable Automotive Dashboards on Liquid Crystal Displays," *Cognition, Technology & Work* 6, no. 4 (2004): 247–65, http://doi.org/10.1007/s10111-004-0163-1.
6 For more, see Kai H. Lim, Izak Benbasat, and Peter A. Todd, "An Experimental Investigation of the Interactive Effects of Interface Style, Instructions, and Task Familiarity on User Performance," *ACM Transactions on Computer-Human Interaction* 3, no. 1 (1996): 1–37, http://doi.org/10.1145/226159.226160.
7 See Lim, "An Experimental."
8 For more, see Ogan M. Yigitbasioglu, and Oana Velcu, "A Review of Dashboards in Performance Management: Implications for Design and Research," *International Journal of Accounting Information Systems* 13, no. 1 (2012): 41–59, http://doi.org/10.1016/j.accinf.2011.08.002.

REFERENCES

1. Rasmussen, Jens. 1983. "Skills, Rules, and Knowledge; Signals, Signs, and Symbols, and Other Distinctions in Human Performance Models." *IEEE Transactions on Systems, Man, and Cybernetics* 13(3): 257–66. doi:10.1109/TSMC.1983.6313160.
2. Reason, J. 1990. *Human Error.* Cambridge: Cambridge University Press.
3. Republished with permission of American Society of Mechanical Engineers, from Corrado, Jonathan. 2022. "Proactive Human Error Reduction Using the Systems Engineering Process." *ASME Journal of Nuclear Engineering and Radiation Science* 8(2). doi:10.1115/1.4051362; permission conveyed through Copyright Clearance Center, Inc.
4. Republished with permission of American Society of Mechanical Engineers, from Corrado, Jonathan. 2022. "The Incorporation of Human Performance Improvement into Systems Design." *ASME Journal of Nuclear Engineering and Radiation Science* 8(2). doi:10.1115/1.4051792; permission conveyed through Copyright Clearance Center, Inc.
5. Hoffman, David. 1999. "I Had a Funny Feeling in My Gut." *Washington Post Foreign Service*, February 10.
6. Corrado, Jonathan. 2017. "Technological Advances, Human Performance, and the Operation of Nuclear Facilities." PhD dissertation, Colorado State University, Fort Collins, CO, ProQuest (AAT 10258407).
7. Republished with permission of American Society of Mechanical Engineers, from Corrado, Jonathan. 2021. "The Intersection of Advancing Technology and Human Performance." *ASME Journal of Nuclear Engineering and Radiation Science* 7(1). doi:10.1115/1.4047717; permission conveyed through Copyright Clearance Center, Inc.
8. Johannsen, Gunnar. 2009. "Human Machine Interaction." In *Control Systems, Robotics and Automation, vol. 21: Elements of Automation*, edited by Heinz D. Unbehauen, 132–62. Oxford: Eolss Publishers.
9. Rogers, Mary Jo. 2013. *Nuclear Energy Leadership: Lessons Learned from U.S. Operators.* Tulsa, OK: PennWell Corp.
10. Republished with permission of the Korean Nuclear Society, from Corrado, Jonathan. 2021. "Human-Machine System Optimization in Nuclear Facility Systems." *Nuclear Engineering and Technology* 53(10). doi:10.1016/j.net.2021.04.022.
11. Republished with permission of American Society of Mechanical Engineers, from Corrado, Jonathan, and Ronald Sega. 2020. "Impact of Advancing Technology on Nuclear Facility Operation." *ASCE-ASME Journal of Risk and Uncertainty in Engineering Systems, Part B: Mechanical Engineering* 6(1). doi:10.1115/1.4044784; permission conveyed through Copyright Clearance Center, Inc.

12. Republished with permission of American Society of Mechanical Engineers, from Corrado, Jonathan. 2022. "Survey Use to Validate Engineering Methodology and Enhance System Safety." *ASCE-ASME Journal of Risk and Uncertainty in Engineering Systems, Part B: Mechanical Engineering* 8(2). doi:10.1115/1.4053305; permission conveyed through Copyright Clearance Center, Inc.
13. Lewis, J., and Sauro, J. 2016. *Qualifying the User Experience. Practical Statistics for User Research.* 2nd ed. Denver, CO: Create Space Publishing.

8 Final Recommendations for the Favorable Combination of Advancing Technology and Human Performance in the Nuclear Setting

RESEARCH RECAP

The systems associated with nuclear facilities are some of the most complex ever developed. As a result, it is important to consider whether the complexity and changing technologies used at nuclear facilities may exacerbate the cost of incidents caused by human error. This book has investigated the relationship

DOI: 10.1201/9781003346265-8

between human performance, technological advances, and the complex systems involved with nuclear facilities.

My primary research question was whether technological advances in the complex systems of nuclear facilities increase the severity, in terms of financial cost, of incidents caused by human error. This is an important question because most nuclear facilities continually update their technology for varying reasons. This continuous technological improvement can frequently require operators to change their routines and their method of interacting with the complex system and the new technology.

To answer this question, hypotheses were established, and a study was conducted to determine whether technological advances affect the interaction between operators and the systems that they operate, resulting in an increased cost of incidents related to human error. The hypothesis was stated as follows:

H_1: Technological advances at a nuclear facility that affect how operators interact within the system do increase the cost of incidents caused by human error.

The null hypothesis was stated as follows:

H_0: Technological advances at a nuclear facility that affect how operators interact within the system do not increase the cost of incidents caused by human error.

CONCLUSIONS REACHED

The *t*-test was used to determine whether there was a statistical difference in the costs of nuclear incidents related to operator error when interacting with a system where recent technological advances have been installed, relative to other incidents. The *t*-test indicated a statistically significant difference between the two groups; therefore, the null hypothesis was rejected. The evidence indicated that, indeed, technological advances at a nuclear facility that affect how operators interact within the system do increase the cost of incidents caused by human error.

As a follow-up study, the question of whether organizations benefit overall by incorporating advanced technology at their facilities was analyzed. The data from this analysis indicated that spending more money on upgrades will increase a facility's capacity factor (i.e., the ratio between observed output and potential output if the facility were consistently running at its full capacity) as well as the number of incidents reported. However, the incidents in the facilities I randomly selected were relatively minor. Given that the nuclear facilities produce vast amounts of power, and the upgrades significantly increase the capacity factor, there appears to be a financial advantage in conducting upgrades, but this benefit should be weighed against the increased rate of Levels 1 and 2 incidents observed.

Based on the information and analysis discussed above, there is evidence that technological advances at a nuclear facility are worth the risk, even though they may increase the cost of resulting incidents due to human error. Due to these findings, additional study was conducted on the impact of human factors (including human error) on plant operation and the phases of the systems engineering process. This was followed by a discussion of the impact of evolving technology on today's facility operators and ways to overcome these challenges by using the systems engineering process.

RECOMMENDATIONS FOR FURTHERING THIS RESEARCH TO BOLSTER THE NUCLEAR INDUSTRY FOR A BRIGHT FUTURE

This study resulted in a rejection of the null hypothesis. Therefore, there is evidence that changing the technology with which operators interact increases the cost of incidents resulting from human error. As a result of this conclusion, I offer three final recommendations.

First, I recommend additional research on this topic. Specifically, a study could be conducted on the rate of small-scale incidents (i.e., those at Level 1 or 2 of the International Nuclear and Radiological Event Scale (INES)). Many of these incidents are not as intricately studied as those that reach a higher level of severity. Instead of an in-depth analysis of each incident, a simple count could be made of the minor incidents that follow changes in technology. A similar procedure to what I use here could be used for data analysis. Instead of using the cost of the incident as the quantitative measure, the number of minor incidents could be used. This approach would eliminate the need to search for cost information on each incident or the details of the incident report other than the INES rating.

My second recommendation concerns the fact that the operator has a unique perspective relevant to the successful operation of the system or component; therefore, their input can result in appreciably less complication after a system or component is installed and operational. It is recommended that systems engineering process programmatic enhancements be considered to tap this resource more effectively and efficiently. I also recommend that research

TABLE 8.1 Minor Events on Each Day during November 2013 [Modified from Corrado, Jonathan. 2017. "Technological Advances, Human Performance, and the Operation of Nuclear Facilities," PhD Dissertation, Colorado State University, Fort Collins, CO, ProQuest (AAT 10258407)]

DAY OF MONTH (NOVEMBER 2013)	*TA*	*NO TA*	*TOTAL # OF EVENTS*
1	2	1	3
4	4	1	5
5	1	2	3
6	3	3	6
7	1	3	4
8	4	1	5
12	3	2	5
13	1	0	1
14	1	1	2
15	3	1	4
18	3	2	5
19	4	2	6
20	3	0	3
21	2	1	3
22	1	2	3
25	2	1	3
26	0	0	0
27	2	1	3
29	1	3	4

be done to quantifiably determine the impact from harnessing operator perspectives in the early stages of the systems engineering process.

I not only recommend but also urge that the systematic and continuous improvement of training, procedures, and human performance be further researched, refined, and evaluated to quantitatively demonstrate a decrease in human error and an increase in plant safety.

Minor Event Study

This work has shown numerous ways these recommendations can be moved forward, but I would like to take a final moment to demonstrate my first recommendation in this chapter that the rate of small incidents at nuclear power plants be investigated because it may seem vague.

As I said, I am specifically talking about incidents rated as Level 1 or 2 on the INES. For this example, I chose November 2013 as there was sufficient data available during this time frame to conduct a thorough analysis. Table 8.1 illustrates the number of events that occurred on each day. They are divided into those associated with technological advances (TA) and those not associated with TA.

I should note that the results for this section were obtained simply by counting the number of incidents reported and determining if any significant technological advances were installed at these facilities in the three months preceding the incident. The incidents were not screened by type and by whether they would be obviously due to operator error. For example, one of the incidents involved the presence of an alcoholic beverage inside the plant near the operations center. Some would argue that this could have nothing to do with technological changes; others might assert that the additional anxiety created by having to deal with technological changes led to the aberrant behavior. Because the hypothesis is that advances increase the likelihood of incidents occurring, this is probably not an important consideration.

It appears that changes to the technology increase the number of minor incidents at Level 1 or 2 on the INES. However, they do not significantly increase the probability of more severe incidents (i.e., those at Level 3 or higher on the INES).

FINAL THOUGHTS

Technological upgrades are unavoidable, so consistent efforts must be devoted toward keeping human error as low as possible amidst those changes. As the evolution of systems progresses, the role of humans within the system will inevitably continue to evolve as well. This evolution necessitates a continued emphasis on human performance improvement as part of maximizing system efficiency and safety. Within this context, as new systems are designed and the human role is transformed, human error should be targeted for substantial reduction relative to predecessor systems, with a corresponding increase in system stability and safety. For this purpose, the roles of humans and machines must be optimized in the design and implementation of new types of automation, but it must be understood that human interaction with systems is compulsory and of major importance [2].

Unfortunately, human error reduction and system design and deployment are often treated as two separate subjects with their own distinct processes that commonly intersect upon the conclusion of design, prior to operation. This traditional approach to system design may have been successful for the obsolete technologies of the past but is proving to be problematic for the design of the more complicated systems of the present. As the complexity of advancing technologies crescendos, human-system interaction warrants a more prominent role in system design and therefore compels early consideration, deliberation, and integration in the beginning stages of the systems engineering process. Incorporation of human error prevention means the system design process harvests the sound development of systems with an improved probability of successful, error-reduced operation [2, 3].

REFERENCES

1. Corrado, Jonathan. 2017. "Technological Advances, Human Performance, and the Operation of Nuclear Facilities." PhD dissertation, Colorado State University, Fort Collins, CO, ProQuest (AAT 10258407).
2. Republished with permission of American Society of Mechanical Engineers, from Corrado, Jonathan, and Ronald Sega. 2020. "Impact of Advancing Technology on Nuclear Facility Operation." *ASCE-ASME Journal of Risk and Uncertainty in Engineering Systems, Part B: Mechanical Engineering* 6(1). doi:10.1115/1.4044784; permission conveyed through Copyright Clearance Center, Inc.
3. Republished with permission of American Society of Mechanical Engineers, from Corrado, Jonathan. 2022. "Proactive Human Error Reduction Using the Systems Engineering Process." *ASME Journal of Nuclear Engineering and Radiation Science* 8(2). doi:10.1115/1.4051362; permission conveyed through Copyright Clearance Center, Inc.

Appendix
Survey Structure, Analysis, and Results

SURVEY DELIVERY METHOD

I used interviews and questionnaires in combination as the survey delivery method. This approach not only obtains quantitative data but also gives respondents the opportunity to express their qualitative thoughts, reactions, and feedback on the subject matter, which will help in assessing the confidence that experts in the engineering field have regarding the proposals in this work.

QUESTION TYPE EMPLOYED

To develop the question set for the survey, I made subsets by discrete subject matter, and questions were written in such a manner as to determine a level of confidence that the concepts would be successful, and that implementation would be feasible. Once the questions had been generated, they were revised based on several relevant concepts fundamental to survey design methodology. I used both closed- and open-ended questions to ensure that the entire breadth of responses could be collected and to provide evidence of expert confidence regarding the concepts proposed in Chapters 6 and 7. For the closed-ended questions, I used a Likert-type scale from 1 to 5 to quantify results on a portion of the survey. Open-ended questions (including the opportunity to comment on answers to closed-ended questions) gave respondents the ability to elaborate on their numerical responses, with the purpose of obtaining evidence that the proposed concepts are sound or identifying where improvement is needed.

DESIGN BIASES CONSIDERED

As indicated above, the initial set of survey questions and iterations thereafter were reviewed relative to several common design biases: leading or loaded questions, overlapping response options, unbalanced response options (including floor and ceiling effects), and framing effects. The survey contained no emotionally charged questions, and the use of Likert-type questions with a neutral response option, along with permitting respondents to elaborate on their answers, made it easier to avoid most of these biases. To avoid framing effects, I used benchmarking to compare the present questions to those from a similar survey [1].

The respondents were advised that anonymity of all information provided on the survey would be maintained.

A pilot study was conducted with three individuals to gauge whether the questions were understandable, consistent, and reliable and that the concepts made sense, as well as to obtain opinions on specific wording and phrasing used [1].

ADDITIONAL CONSIDERATIONS

Additional factors taken into consideration in question development were the avoidance of jargon, slang, and abbreviations in question content; ensuring that questions were written in such a way as to test hypotheses (not questions written *about* hypotheses); maintenance of realistic expectations as to respondent capabilities; and avoidance of negatively phrased questions [1].

Lastly, question ordering was accomplished by taking into consideration organizational concerns (e.g., choice of opening and closing questions, and smooth survey flow) and order effects (content relationships, contextual effects, and rating dependencies). Parceling Chapters 6 and 7 content into manageable sections enabled the respondent to read the information on each specific topic and then answer a series of survey questions on that topic. This ensured that the respondent would not have to presume what the question was referring to and prevented frustration by not compelling the respondent to reread sections [1].

SURVEY POPULATION AND PERSONNEL SELECTION

The survey sample consisted of forty system design engineers (age twenty-five to sixty-five) of various backgrounds and disciplines (e.g., nuclear engineers, electrical engineers, mechanical engineers) who were knowledgeable regarding the systems engineering and design process [1].

To select the personnel for the study, the engineering staff roster of my place of employment was first filtered to exclude staff who did not hold an engineering degree and those who had not been involved in systems engineering or design for at least five years. From this reduced listing, random selection occurred using a random number generator on Microsoft Excel. Thirty individuals were selected using this method, and then additional individuals who had expressed interest were added to the final sample. A solicitation letter, disseminated by email, included a brief explanation of the study, the estimated time required to complete the survey interview, and how the information would be handled and addressed within the study [1].

SURVEY ADMINISTRATION

The respondents were provided with the chapter content and the questions and statements along with the five-point Likert scale. Depending on the question, the potential responses were labeled in one of two ways: (1) "never," "occasionally," "regularly," "frequently," and "always," or (2) "strongly disagree," "disagree," "neither agree or disagree," "agree," and "strongly agree." Using two different scales not only allows for flexibility in questioning but also offers survey range and depth instead of repetition (see Tables A.1–A.6).

In addition, as discussed above, the respondents could make open-ended comments on any of the closed-ended questions, and those comments were recorded in the appropriate box on the survey [1].

TABLE A.1 Operator Involvement in the Systems Engineering Process [Reproduced with Permission of American Society of Mechanical Engineers, from "Survey Use to Validate Engineering Methodology and Enhance System Safety," *ASCE-ASME Journal of Risk and Uncertainty in Engineering Systems, Part B: Mechanical Engineering*, Corrado, Jonathan, 8(2), 2022; Permission Conveyed through Copyright Clearance Center, Inc.]

USING A SCALE OF 1–5 (1 = STRONGLY DISAGREE, 2 = DISAGREE, 3 = NEITHER AGREE OR DISAGREE, 4 = AGREE, 5 = STRONGLY AGREE). PLEASE INDICATE YOUR AGREEMENT WITH THE FOLLOWING CONCEPTS:

#	*STATEMENT*	*STRONGLY DISAGREE*	*DISAGREE*	*NEITHER AGREE OR DISAGREE*	*AGREE*	*STRONGLY AGREE*
1	Operations personnel should be involved in the systems engineering process from the onset.	1	2	3	4	5
2	Because system use encompasses a large portion of the systems life, operations personnel participation in system design is necessary not only for system operational success and support but also for human error reduction in the operation of new technologies.	1	2	3	4	5
3	FMEA can and should be used during all stages of the systems engineering process.	1	2	3	4	5
4	Operations personnel should play a role in FMEA during the design stages.	1	2	3	4	5

TABLE A.1 (*Continued*)

#	*STATEMENT*	*STRONGLY DISAGREE*	*DISAGREE*	*NEITHER AGREE OR DISAGREE*	*AGREE*	*STRONGLY AGREE*
5	Design criteria can come and should come from a number of diverse sources, but those that originate from the plant operations staff should be given elevated deliberation because they will become the system owner through the utilization phase of the system's life.	1	2	3	4	5

USING A SCALE OF 1–5 (1 = NEVER, 2 = OCCASIONALLY, 3 = REGULARLY, 4 = FREQUENTLY, 5 = ALWAYS), PLEASE INDICATE YOUR LIKELIHOOD TO IMPLEMENT OR SUPPORT THE FOLLOWING CONCEPTS:

#	*STATEMENT*	*NEVER*	*OCCASIONALLY*	*REGULARLY*	*FREQUENTLY*	*ALWAYS*
6	Involve operations personnel in the systems engineering process from the onset.	1	2	3	4	5
7	Utilize FMEA techniques during all stages of the systems engineering process.	1	2	3	4	5
8	Involve operations personnel in FMEA during the design stages of the systems engineering process.	1	2	3	4	5
9	Give design criteria that originate from operations personnel elevated deliberation in system design.	1	2	3	4	5

TABLE A.2 Human Performance Association with System Operational Requirements and System Test, Evaluation, and Validation [Reproduced with Permission of American Society of Mechanical Engineers, from "Survey Use to Validate Engineering Methodology and Enhance System Safety," *ASCE-ASME Journal of Risk and Uncertainty in Engineering Systems, Part B: Mechanical Engineering*, Corrado, Jonathan, 8(2), 2022; Permission Conveyed through Copyright Clearance Center, Inc.]

#	*STATEMENT*	*STRONGLY DISAGREE*	*DISAGREE*	*NEITHER AGREE OR DISAGREE*	*AGREE*	*STRONGLY AGREE*
10	Embedded within the elements of system operational requirements should be specific, exclusive human performance system requirements that can be easily assessed and discernable in system design.	1	2	3	4	5
11	The creation of human performance TPMs that evaluate the integration of the human error reduction tools with the technology necessary to achieve functionality required for system purpose should be given sufficient attention.	1	2	3	4	5
12	Error precursors require systematic evaluation, logical selection or generation, and potential modification before adaptation into human performance TPMs.	1	2	3	4	5

TABLE A.2 *Continued*

USING A SCALE OF 1–5 (1 = NEVER, 2 = OCCASIONALLY, 3 = REGULARLY, 4 = FREQUENTLY, 5 = ALWAYS), PLEASE INDICATE YOUR LIKELIHOOD TO IMPLEMENT OR SUPPORT THE FOLLOWING CONCEPTS:

#	*STATEMENT*	*NEVER*	*OCCASIONALLY*	*REGULARLY*	*FREQUENTLY*	*ALWAYS*
13	Establish and incorporate human performance system operational requirements into system design.	1	2	3	4	5
14	Establish and incorporate human performance TPMs that evaluate the integration of the human error reduction tools with the technology necessary to achieve functionality required for system purpose.	1	2	3	4	5

TABLE A.3 Procedure and Training Development in the Systems Engineering Process [Reproduced with Permission of American Society of Mechanical Engineers, from "Survey Use to Validate Engineering Methodology and Enhance System Safety," *ASCE-ASME Journal of Risk and Uncertainty in Engineering Systems, Part B: Mechanical Engineering*, Corrado, Jonathan, 8(2), 2022; Permission Conveyed through Copyright Clearance Center, Inc.]

USING A SCALE OF 1–5 (1 = STRONGLY DISAGREE, 2 = DISAGREE, 3 = NEITHER AGREE OR DISAGREE, 4 = AGREE, 5 = STRONGLY AGREE), PLEASE INDICATE YOUR AGREEMENT WITH THE FOLLOWING CONCEPTS:

#	*STATEMENT*	*STRONGLY DISAGREE*	*DISAGREE*	*NEITHER AGREE OR DISAGREE*	*AGREE*	*STRONGLY AGREE*
15	The generation of operations procedures should begin at the conceptual design stage in the systems engineering process and then further developed and refined as the systems engineering process advances.	1	2	3	4	5
16	Proper application of human error prevention tools and techniques should be soundly intertwined into the framework of the operations procedures.	1	2	3	4	5
17	In the generation of operations procedures, every effort should be made to drive human performance into the skill-based mode.	1	2	3	4	5
18	When manipulating technologically advanced, complex systems, operators should be trained using the knowledge-based training approach to ensure adherence to design boundaries, efficiency in operation, and an adequate margin to safety.	1	2	3	4	5

TABLE A.3 Continued

#	*USING A SCALE OF 1–5 (1 = STRONGLY DISAGREE, 2 = DISAGREE, 3 = NEITHER AGREE OR DISAGREE, 4 = AGREE, 5 = STRONGLY AGREE), PLEASE INDICATE YOUR AGREEMENT WITH THE FOLLOWING CONCEPTS:* STATEMENT	*STRONGLY DISAGREE*	*DISAGREE*	*NEITHER AGREE OR DISAGREE*	*AGREE*	*STRONGLY AGREE*
19	Operator training generation should begin at conceptual design and be further developed as the systems engineering process advances.	1	2	3	4	5
20	Included in a robust knowledge-based training program should be a vigorous plant casualty control drill program.	1	2	3	4	5
21	The drill program involving the system in design should be developed during the conceptual design stage and iterated as the systems engineering process progresses.	1	2	3	4	5
22	Drill scenario development can provide another avenue for system design review and evaluation by, potentially, an additional set of eyes viewing it from an alternant perspective.	1	2	3	4	5
23	The development of the drill program during the conceptual design stage and iteration through the remainder of the systems engineering process could provide feedback to system designers for desirable improvements and cultivates the continued development of operator training and procedures.	1	2	3	4	5

(Continued)

TABLE A.3 ***Continued***

USING A SCALE OF 1–5 (1 = NEVER, 2 = OCCASIONALLY, 3 = REGULARLY, 4 = FREQUENTLY, 5 = ALWAYS), PLEASE INDICATE YOUR LIKELIHOOD TO IMPLEMENT OR SUPPORT THE FOLLOWING CONCEPTS:						
#	*STATEMENT*	*NEVER*	*OCCASIONALLY*	*REGULARLY*	*FREQUENTLY*	*ALWAYS*
24	Generate operations procedures during the conceptual design stage in the systems engineering process further develop and refine the procedures as the systems engineering process advances.	1	2	3	4	5
25	Intertwine human error prevention tools and techniques into the framework of operations procedures.	1	2	3	4	5
26	Train operators using the knowledge-based training approach to ensure adherence to design boundaries, efficiency in operation, and an adequate margin to safety.	1	2	3	4	5
27	Generate operator training during the conceptual design stage and further develop the training as the systems engineering process advances.	1	2	3	4	5
28	Include a vigorous plant casualty control drill program within a robust knowledge-based training program.	1	2	3	4	5
29	Generate a drill program during the conceptual design stage and further develop the training as the systems engineering process advances.	1	2	3	4	5

TABLE A.4 Operator Attribute Determination [Reproduced with Permission of American Society of Mechanical Engineers, from "Survey Use to Validate Engineering Methodology and Enhance System Safety," *ASCE-ASME Journal of Risk and Uncertainty in Engineering Systems, Part B: Mechanical Engineering*, Corrado, Jonathan, 8(2), 2022; Permission Conveyed through Copyright Clearance Center, Inc.]

#	*STATEMENT*	*STRONGLY DISAGREE*	*DISAGREE*	*NEITHER AGREE OR DISAGREE*	*AGREE*	*STRONGLY AGREE*
30	The determination and development of necessary operator skills and training requirements should begin at the conceptual design stage in the systems engineering process and be refined as the systems engineering process advances.	1	2	3	4	5
31	Determined operator attributes can manifest as system requirements determined during requirements analysis.	1	2	3	4	5
32	Workforce planning should be built in all stages of the systems engineering process to ensure personnel with the necessary attributes are available when the system is deployed.	1	2	3	4	5
33	The development of the human should be as important as the design of the system.	1	2	3	4	5

(Continued)

TABLE A.4 ***Continued***

#	*STATEMENT*	*STRONGLY DISAGREE*	*DISAGREE*	*NEITHER AGREE OR DISAGREE*	*AGREE*	*STRONGLY AGREE*
34	As the systems engineering process progresses, management needs to be constantly assessing the connection between leadership practices, employee work passion, customer devotion, and the bottom line.	1	2	3	4	5
35	There is a clear connection between the quality of an organization's leadership practices, as perceived by employees, and subsequent intentions by personnel to stay with an organization, perform at a high level, and apply discretionary effort.	1	2	3	4	5

USING A SCALE OF 1–5 (1 = NEVER, 2 = OCCASIONALLY, 3 = REGULARLY, 4 = FREQUENTLY, 5 = ALWAYS), PLEASE INDICATE YOUR LIKELIHOOD OF IMPLEMENT OR SUPPORT THE FOLLOWING CONCEPTS:

#	*STATEMENT*	*NEVER*	*OCCASIONALLY*	*REGULARLY*	*FREQUENTLY*	*ALWAYS*
36	Determine and develop necessary operator skills and training requirements at the conceptual design stage in the systems engineering process and refine them as the systems engineering process advances.	1	2	3	4	5

TABLE A.4 *Continued*

USING A SCALE OF 1–5 (1 = NEVER, 2 = OCCASIONALLY, 3 = REGULARLY, 4 = FREQUENTLY, 5 = ALWAYS), PLEASE INDICATE YOUR LIKELIHOOD OF IMPLEMENT OR SUPPORT THE FOLLOWING CONCEPTS:

#	*STATEMENT*	*NEVER*	*OCCASIONALLY*	*REGULARLY*	*FREQUENTLY*	*ALWAYS*
37	Build workforce planning into all stages of the systems engineering process to ensure personnel with the necessary attributes are available when the system is deployed.	1	2	3	4	5
38	Develop the human operator with the same rigor and attention as the design of the system.	1	2	3	4	5
39	Constantly assessing the connection between leadership practices, employee work passion, customer devotion, and the bottom line as the systems engineering process progresses.	1	2	3	4	5

TABLE A.5 Systems Engineering Infrastructure [Reproduced with Permission of American Society of Mechanical Engineers, from "Survey Use to Validate Engineering Methodology and Enhance System Safety," *ASCE-ASME Journal of Risk and Uncertainty in Engineering Systems, Part B: Mechanical Engineering*, Corrado, Jonathan, 8(2), 2022; Permission Conveyed through Copyright Clearance Center, Inc.]

#	*STATEMENT*	*STRONGLY DISAGREE*	*DISAGREE*	*NEITHER AGREE OR DISAGREE*	*AGREE*	*STRONGLY AGREE*
40	The design and development of a system necessitates an adaptive and unique systems engineering infrastructure.	1	2	3	4	5
41	With respect to systems engineering infrastructure, it is important to remain apprised and parallel to the industry standard and best practices regarding usage of organizational design, management processes, software development and functionality, and administrative methods.	1	2	3	4	5
42	As the systems engineering process for product or system development is launched, the proper infrastructure should be established to efficiently support the system development.	1	2	3	4	5
43	The infrastructure design should be a formal, guided process for integrating the people, information, and technology of an organization.	1	2	3	4	5

TABLE A.5 ***Continued***

USING A SCALE OF 1–5 (1 = NEVER, 2 = OCCASIONALLY, 3 = REGULARLY, 4 = FREQUENTLY, 5 = ALWAYS), PLEASE INDICATE YOUR LIKELIHOOD TO IMPLEMENT OR SUPPORT THE FOLLOWING CONCEPTS:						
#	*STATEMENT*	*NEVER*	*OCCASIONALLY*	*REGULARLY*	*FREQUENTLY*	*ALWAYS*
44	Establish the proper infrastructure to efficiently support the system development as the systems engineering process for product or system development is launched.	1	2	3	4	5

TABLE A.6 Means to Minimize Human Error Impact throughout the Systems Engineering Process [Modified from Corrado, Jonathan. 2017. "Technological Advances, Human Performance, and the Operation of Nuclear Facilities," PhD Dissertation, Colorado State University, Fort Collins, CO, ProQuest (AAT 10258407)]

USING A SCALE OF 1–5 (1 = STRONGLY DISAGREE, 2 = DISAGREE, 3 = NEITHER AGREE OR DISAGREE, 4 = AGREE, 5 = STRONGLY AGREE), PLEASE INDICATE YOUR AGREEMENT WITH THE FOLLOWING CONCEPTS:

#	*STATEMENT*	*STRONGLY DISAGREE*	*DISAGREE*	*NEITHER AGREE OR DISAGREE*	*AGREE*	*STRONGLY AGREE*
45	An HPI process should be established not only to remain cognizant of human performance during system design but to provide a means to evaluate human performance of the system at the various stages.	1	2	3	4	5
46	The HPI process can and should be established early in the systems engineering process to identify the impact and extent to which human error affects plant systems and equipment, which will provide necessary feedback into system design process.	1	2	3	4	5
47	To institute the HPI process in a manageable fashion, organizations should determine human error severity criteria and establish a tracking system to capture these human error-induced potential issues for engineered and administrative component features, human error prevention training and reinforcement, and lessons learned.	1	2	3	4	5

TABLE A.6 Continued

USING A SCALE OF 1–5 (1 = STRONGLY DISAGREE, 2 = DISAGREE, 3 = NEITHER AGREE OR DISAGREE, 4 = AGREE, 5 = STRONGLY AGREE), PLEASE INDICATE YOUR AGREEMENT WITH THE FOLLOWING CONCEPTS:

#	STATEMENT	STRONGLY DISAGREE	DISAGREE	NEITHER AGREE OR DISAGREE	AGREE	STRONGLY AGREE
48	These potential issues and incidents can be discovered during the many steps of each stage of the systems engineering process, during the development of procedures for system operation, during training development, during FMEA, and during the design reviews conducted at the conclusion of the conceptual, preliminary, and detail design and development stages of the systems engineering process.	1	2	3	4	5
49	As a part of the HPI process, organizations should create a formal process to examine incidents and perceived vulnerabilities that occur throughout the development of the system during the systems engineering process and, later, during plant operation.	1	2	3	4	5
50	As a part of the HPI process, a methodical approach should be established to systematically direct the potential incident or vulnerability through a series of questions and facilitators to determine regulatory impact, safety significance, determine which organization will perform the corrective action or causal analysis, and ultimately correct the deficiency.	1	2	3	4	5

(Continued)

TABLE A.6 ***Continued***

USING A SCALE OF 1–5 (1 = STRONGLY DISAGREE, 2 = DISAGREE, 3 = NEITHER AGREE OR DISAGREE, 4 = AGREE, 5 = STRONGLY AGREE), PLEASE INDICATE YOUR AGREEMENT WITH THE FOLLOWING CONCEPTS:

#	*STATEMENT*	*STRONGLY DISAGREE*	*DISAGREE*	*NEITHER AGREE OR DISAGREE*	*AGREE*	*STRONGLY AGREE*
51	Due diligence, attention, constant emphasis, and management support needs to be given to the identification and reporting of the potential human errors and the corrective action processing of the problems as they are reported	1	2	3	4	5
52	This process, if effectively performed, will constantly reinforce good human performance practice, the reduction of human error in system design and potential human errors during operation, proportionally decrease the volume of potential problems at all levels, and continually drive errors to the least significant level.	1	2	3	4	5
53	A management team should be established to help ensure that the HPI program is effectively and consistently implemented with particular emphasis on evaluating significant issues, adverse trends identified, and ineffective corrective actions that were applied to conditions adverse to safety and security.	1	2	3	4	5

TABLE A.6 ***Continued***

USING A SCALE OF 1–5 (1 = STRONGLY DISAGREE, 2 = DISAGREE, 3 = NEITHER AGREE OR DISAGREE, 4 = AGREE, 5 = STRONGLY AGREE), PLEASE INDICATE YOUR AGREEMENT WITH THE FOLLOWING CONCEPTS:

#	*STATEMENT*	*STRONGLY DISAGREE*	*DISAGREE*	*NEITHER AGREE OR DISAGREE*	*AGREE*	*STRONGLY AGREE*
54	Management should promote a culture focused on safety and operational excellence where employees should be not only encouraged to make error incident reports but also expected to make these reports in a retaliation-free atmosphere.	1	2	3	4	5

USING A SCALE OF 1–5 (1 = NEVER, 2 = OCCASIONALLY, 3 = REGULARLY, 4 = FREQUENTLY, 5 = ALWAYS), PLEASE INDICATE YOUR LIKELIHOOD TO IMPLEMENT OR SUPPORT THE FOLLOWING CONCEPTS:

#	*STATEMENT*	*NEVER*	*OCCASIONALLY*	*REGULARLY*	*FREQUENTLY*	*ALWAYS*
55	Establish an HPI process to not only to remain cognizant of human performance during system design but also to provide a means to evaluate human performance of the system at the various stages.	1	2	3	4	5
56	Establish an HPI process early in the systems engineering process to identify the impact and extent to which human error affects plant systems and equipment, which will provide necessary feedback into system design process.	1	2	3	4	5

(Continued)

TABLE A.6 ***Continued***

USING A SCALE OF 1–5 (1 = NEVER, 2 = OCCASIONALLY, 3 = REGULARLY, 4 = FREQUENTLY, 5 = ALWAYS), PLEASE INDICATE YOUR LIKELIHOOD TO IMPLEMENT OR SUPPORT THE FOLLOWING CONCEPTS:						
#	*STATEMENT*	*NEVER*	*OCCASIONALLY*	*REGULARLY*	*FREQUENTLY*	*ALWAYS*
57	Determine human error severity criteria and establish a tracking system to capture these human error-induced potential issues for engineered and administrative component features, human error prevention training and reinforcement, and lessons learned.	1	2	3	4	5
58	Create a formal process to examine incidents and perceived vulnerabilities that occur throughout the development of the system during the systems engineering process and, later, during plant operation.	1	2	3	4	5
59	Establish a methodical approach to systematically direct the potential incident or vulnerability through a series of questions and facilitators to determine regulatory impact, safety significance, determine which organization will perform the corrective action or causal analysis, and ultimately correct the deficiency.	1	2	3	4	5

TABLE A.6 ***Continued***

USING A SCALE OF 1–5 (1 = NEVER, 2 = OCCASIONALLY, 3 = REGULARLY, 4 = FREQUENTLY, 5 = ALWAYS), PLEASE INDICATE YOUR LIKELIHOOD TO IMPLEMENT OR SUPPORT THE FOLLOWING CONCEPTS:

#	*STATEMENT*	*NEVER*	*OCCASIONALLY*	*REGULARLY*	*FREQUENTLY*	*ALWAYS*
60	Establish a management team to help ensure that the HPI program is effectively and consistently implemented with particular emphasis on evaluating significant issues, adverse trends identified, and ineffective corrective actions that were applied to conditions adverse to safety and security.	1	2	3	4	5
61	Promote a culture focused on safety and operational excellence where employees should be not only encouraged to make error incident reports but also expected to make these reports in a retaliation-free atmosphere.	1	2	3	4	5

HUMAN SUBJECT PROTECTIONS

All research participants signed a written consent form explaining their rights as a volunteer, the ability to withdraw their consent at any time during the study, the fact that no identifying information was collected or used in the analysis of the results, and reasonable assurance that the researchers had taken reasonable safeguards to minimize any known or potential but unknown risks and that there were no known risks [1].

OVERALL ANALYSIS OF SURVEY RESULTS

Ideally, responses can be compared to an industry benchmark, a competitor, or even a similar question from a prior survey. In most cases, however, such data do not exist because they are too expensive or too difficult to obtain. Because the questions in this survey were original to this study, there are no historical or comparative data. The following quantitative analyses were performed using the data from Table A.7:

1. **Percentage agreeing**: The percentage of respondents who indicated "agree" or "strongly agree" (or "frequently" or "always") on an item.
    a. Analysis yielded a score of 96.6 percent for the percentage agreeing
2. **Top box and top two box scoring**: The top box refers to responses of "strongly agree" or "always." In this case of this data set and analysis, the top two box score is the same as the score reported in result 1 above.
    a. Analysis yielded a score of 96.6 percent for top two box
    b. Analysis yielded a score of 63.1 percent for top box
3. **Net top box**: Found by counting the number of respondents selecting the top choice and subtracting the number who selected the bottom choice.
    a. Analysis yielded a score of 63.1 percent for net top box
4. ***z*-Score to Percentile Rank**: This converts the raw score into a normal score because rating scale means often follow a normal or close to normal distribution. Eighty percent of the number of points in a scale is a reasonable benchmark to compare the mean to. For this analysis, 4 is used ($5 \times 0.80 = 4$). First, subtract the benchmark from the mean.

TABLE A.7 Survey Data [Reproduced with Permission of American Society of Mechanical Engineers, from "Survey Use to Validate Engineering Methodology and Enhance System Safety," *ASCE-ASME Journal of Risk and Uncertainty in Engineering Systems, Part B: Mechanical Engineering*, Corrado, Jonathan, 8(2), 2022; Permission Conveyed through Copyright Clearance Center, Inc.]

Q\N	*1*	*2*	*3*	*4*	*5*	*6*	*7*	*8*	*9*	*10*	*11*	*12*	*13*	*14*	*15*	*16*	*17*	*18*	*19*	*20*	*21*	*22*	*23*	*24*	*25*	*26*	*27*	*28*	*29*	*30*	*31*	*32*	*33*	*34*	*35*	*36*	*37*	*38*	*39*	*40*
1	5	5	5	5	5	4	5	5	5	5	5	4	5	5	5	5	5	4	5	5	5	4	5	5	4	5	5	4	4	4	5	4	5	5	4	4	5	4	5	5
2	5	4	5	4	5	5	5	5	5	5	5	5	5	3	5	5	5	5	5	5	5	5	4	5	5	5	5	5	5	5	4	5	4	5	5	5	4	5	4	5
3	4	5	4	5	4	5	4	4	5	4	4	5	4	4	4	4	4	5	4	4	4	5	5	4	4	5	5	4	5	4	5	5	5	5	5	5	5	4	5	4
4	4	4	5	3	4	4	5	5	4	3	5	4	5	4	5	5	4	4	5	4	5	4	5	4	5	4	4	5	5	5	4	4	5	5	4	4	5	5	5	4
5	5	5	5	5	5	4	5	5	5	5	5	5	5	5	5	5	5	5	5	5	4	5	4	5	4	5	5	4	5	5	5	5	3	4	4	4	3	5	4	5
6	5	4	5	5	4	5	4	5	5	4	4	5	5	4	4	5	5	5	4	5	4	4	5	4	5	5	4	4	4	4	5	4	5	5	5	5	4	5	5	4
7	5	5	4	5	5	5	5	4	3	5	5	5	5	5	3	5	4	5	5	5	5	5	5	4	4	3	5	5	3	5	4	5	3	4	4	4	5	4	4	4
8	4	5	4	5	4	5	4	4	4	4	4	4	4	5	4	4	4	4	4	4	5	4	4	4	5	4	5	4	4	5	4	4	4	5	5	4	5	5	5	5
9	5	5	5	4	5	5	5	4	5	5	5	5	5	4	5	5	5	5	5	5	4	5	5	5	4	4	5	5	5	4	5	5	5	5	5	3	5	5	4	4
10	5	5	5	5	5	5	5	5	5	5	5	5	5	5	5	5	5	5	5	5	5	4	4	4	5	5	4	4	5	3	5	4	5	5	5	4	5	4	5	5
11	4	4	5	4	5	4	5	4	5	5	4	5	4	5	4	4	5	4	5	5	4	4	5	4	4	5	5	5	5	4	4	4	4	5	4	4	4	4	5	5
12	5	5	4	5	5	4	5	4	4	5	4	5	5	4	5	5	5	5	5	5	4	5	4	5	5	3	5	5	5	5	5	5	5	5	4	5	3	5	4	4
13	5	4	5	4	5	5	4	5	5	5	5	4	5	5	5	5	4	5	5	5	4	4	5	4	4	5	5	4	4	5	3	5	5	5	4	4	5	3	5	4
14	4	4	5	4	4	5	4	4	4	5	4	4	4	5	4	4	5	5	4	4	5	5	5	3	4	5	5	5	5	4	5	4	4	4	3	5	4	5	4	5
15	5	5	5	5	5	4	5	5	5	5	5	4	5	5	5	5	5	4	5	5	4	5	5	4	5	4	4	4	4	5	4	4	5	5	4	4	5	4	4	4
16	5	5	4	4	5	5	4	5	4	4	5	4	4	4	5	5	4	4	5	4	5	4	4	5	4	5	5	5	5	5	4	5	5	4	5	4	5	5	5	5
17	5	5	5	5	5	5	3	5	5	5	5	5	5	5	5	5	5	5	5	5	5	4	5	4	4	5	4	4	4	4	5	3	4	5	5	5	5	5	4	4
18	5	5	4	5	5	5	5	5	5	5	5	5	5	5	5	5	5	5	5	5	5	5	4	3	5	4	5	5	5	3	4	5	5	5	3	5	4	4	5	5
19	4	4	3	5	4	5	4	4	5	5	4	5	3	4	5	4	5	4	4	5	4	5	5	4	5	4	5	4	5	4	5	3	5	5	4	4	5	5	5	4
20	5	5	5	5	5	5	4	5	4	5	4	5	5	4	5	5	4	5	5	5	4	5	4	4	4	3	5	5	5	5	4	5	5	5	5	4	5	4	4	5

(Continued)

TABLE A.7 ***Continued***

Q\N	1	2	3	4	5	6	7	8	9	10	11	12	13	14	15	16	17	18	19	20	21	22	23	24	25	26	27	28	29	30	31	32	33	34	35	36	37	38	39	40
21	5	4	5	5	5	4	5	4	5	4	5	4	4	5	4	5	5	5	5	5	5	5	5	5	4	5	4	5	4	5	4	4	4	5	3	5	4	4	5	4
22	5	5	5	5	5	5	3	5	5	5	5	5	5	5	5	5	5	5	5	5	4	4	3	5	5	5	5	4	5	5	5	4	5	5	4	5	5	5	5	5
23	4	4	5	4	4	4	5	4	5	4	5	4	5	4	4	4	5	4	4	4	4	5	5	4	4	4	5	5	4	5	5	3	4	4	5	4	5	4	4	4
24	5	4	4	5	4	5	4	4	4	4	4	5	4	4	5	5	4	4	4	5	5	5	5	5	5	5	4	4	5	4	4	4	5	5	4	5	4	5	5	4
25	5	5	5	5	5	5	5	5	5	5	5	5	5	5	5	5	5	5	5	5	4	5	4	4	4	4	5	5	3	5	5	5	4	5	5	5	3	5	4	4
26	5	5	5	3	5	5	5	5	5	5	5	5	5	5	5	5	5	5	5	5	4	4	5	5	4	4	5	4	4	5	5	4	5	5	4	4	5	4	5	5
27	5	5	4	5	5	4	5	5	4	4	5	4	4	5	5	5	5	5	5	5	5	5	4	3	5	5	4	5	5	4	5	5	5	5	4	5	5	4	5	5
28	5	5	5	5	4	5	5	4	5	5	4	5	5	4	5	5	5	5	5	5	5	4	5	5	3	5	3	3	5	5	4	5	4	5	5	4	4	5	4	5
29	4	4	5	4	5	5	4	4	5	4	4	5	5	5	4	4	5	5	4	4	5	5	4	4	4	4	5	5	5	4	4	4	5	5	4	5	5	5	5	4
30	5	5	5	5	5	5	5	5	5	5	5	5	5	5	5	5	5	5	5	5	5	4	5	5	5	5	4	4	4	5	3	5	5	4	5	4	4	4	4	5
31	5	5	4	5	5	5	5	5	5	5	5	4	5	5	5	4	5	5	5	5	4	5	5	5	4	4	5	5	5	5	5	4	5	4	4	5	5	4	5	4
32	5	5	5	3	5	5	5	5	5	5	5	5	5	5	5	5	3	5	5	5	5	4	4	4	5	5	4	4	4	5	4	5	4	5	5	5	5	3	4	4
33	4	5	4	5	5	4	5	4	4	4	3	5	4	5	4	4	4	5	4	5	4	5	5	4	4	4	3	3	5	5	5	5	5	5	4	4	5	5	3	4
34	5	5	5	5	5	5	5	5	5	5	5	5	5	5	5	5	5	5	5	4	5	5	4	5	5	4	5	4	5	4	5	5	4	5	5	5	4	4	5	5
35	5	5	5	4	5	5	4	3	5	5	5	5	5	5	5	5	4	5	5	5	5	5	4	4	4	5	5	5	4	5	4	4	5	4	4	5	5	4	5	4
36	5	5	5	5	5	5	5	5	5	5	5	5	5	5	3	5	5	5	5	5	4	5	5	4	4	4	4	4	5	5	5	5	5	5	4	4	4	5	4	4
37	4	4	5	4	4	5	4	4	4	5	4	4	5	4	4	4	5	4	5	4	5	5	5	5	4	5	5	5	3	5	4	3	5	4	4	5	3	5	5	5
38	4	4	4	5	4	4	4	5	4	4	5	5	4	5	4	5	4	5	4	4	4	5	4	3	5	4	4	4	4	5	5	4	3	5	4	5	5	3	4	4
39	5	5	5	5	5	5	4	5	5	5	5	5	5	5	5	5	5	5	5	5	4	4	5	4	5	5	5	5	5	4	5	5	3	3	5	5	4	4	5	5
40	5	5	5	5	5	5	5	5	5	5	5	5	5	5	5	5	5	5	5	5	5	5	4	5	4	4	4	5	4	5	4	5	4	5	4	4	5	4	4	4
41	5	4	5	5	5	5	5	5	3	5	5	5	5	3	5	5	5	5	5	4	4	5	4	4	4	5	5	5	3	5	5	4	5	4	5	4	4	5	5	4
42	4	4	5	4	5	4	5	4	4	4	4	5	5	5	4	5	4	4	5	5	5	4	5	5	5	5	5	4	4	4	4	4	4	4	4	3	5	5	4	5
43	5	5	5	5	5	5	5	5	5	5	5	3	5	5	5	5	5	5	5	5	4	4	4	4	4	4	5	5	5	5	5	5	3	4	4	4	4	4	5	4

TABLE A.7 ***Continued***

Q\N	*1*	*2*	*3*	*4*	*5*	*6*	*7*	*8*	*9*	*10*	*11*	*12*	*13*	*14*	*15*	*16*	*17*	*18*	*19*	*20*	*21*	*22*	*23*	*24*	*25*	*26*	*27*	*28*	*29*	*30*	*31*	*32*	*33*	*34*	*35*	*36*	*37*	*38*	*39*	*40*
44	5	5	5	5	5	5	5	4	5	5	5	5	5	5	5	5	5	5	5	5	5	5	5	5	4	5	5	5	5	5	4	4	4	5	4	5	4	4	4	4
45	5	5	5	5	5	5	5	5	5	5	5	5	5	5	5	5	5	5	5	5	5	5	5	5	5	4	4	4	5	4	5	5	5	5	4	5	5	4	4	4
46	5	5	5	5	5	5	5	5	5	5	5	5	5	5	5	5	5	5	5	5	4	4	4	4	4	5	5	5	4	5	4	4	4	4	5	4	5	5	5	5
47	4	5	4	5	4	4	4	5	5	4	4	5	5	4	4	5	4	5	4	4	5	5	5	5	5	4	4	4	5	5	5	5	4	5	4	5	5	5	4	4
48	5	5	5	3	5	5	5	5	5	5	5	5	3	5	5	5	5	5	5	5	5	5	4	4	4	5	3	5	4	5	4	5	5	5	4	5	4	4	4	4
49	5	5	5	5	5	5	5	5	5	5	5	5	5	5	5	5	5	5	5	5	4	5	5	5	5	5	5	4	5	4	5	4	5	4	4	4	5	4	5	5
50	5	5	5	5	5	5	5	5	5	5	5	5	5	5	5	5	5	5	5	5	4	4	4	5	5	4	5	5	5	5	5	5	5	5	5	3	4	5	5	4
51	5	5	5	5	5	5	5	5	5	5	5	5	5	5	5	5	5	5	5	5	5	4	5	5	4	5	5	4	5	5	4	4	4	4	4	4	5	5	4	5
52	4	4	5	4	4	5	4	4	3	5	5	4	4	5	5	4	4	4	5	4	4	3	4	4	5	4	5	5	4	5	4	5	5	4	5	5	5	5	5	4
53	5	5	5	5	5	5	5	5	5	5	5	5	5	5	5	5	5	5	5	5	5	4	4	5	5	4	5	3	3	4	4	4	4	4	4	4	5	4	4	4
54	5	5	5	5	5	5	5	5	5	5	5	5	5	5	5	5	5	5	5	5	5	5	3	4	4	5	5	4	4	5	5	5	3	5	5	3	5	5	5	5
55	5	5	5	5	5	5	5	5	5	5	5	5	5	5	5	5	5	5	5	5	4	5	5	4	5	4	5	5	5	4	4	4	5	5	5	4	3	4	5	4
56	4	4	5	4	4	4	4	4	5	4	4	4	4	4	4	5	4	4	5	4	5	5	5	4	5	4	3	5	5	5	5	5	4	5	5	5	5	5	4	5
57	5	4	4	5	5	4	5	5	4	4	5	4	4	5	4	4	4	5	4	5	5	4	3	4	5	5	3	5	5	4	4	4	3	4	4	4	5	3	5	4
58	4	4	4	4	4	5	4	4	4	4	4	5	4	4	4	4	5	3	4	4	5	5	5	3	5	4	4	4	4	5	5	5	4	5	5	5	4	4	5	5
59	5	5	5	5	5	5	5	5	5	5	5	5	5	5	5	5	5	5	5	5	4	5	5	5	4	5	5	5	5	5	5	4	5	5	5	5	5	4	5	5
60	5	5	4	5	5	4	4	5	5	5	4	5	5	4	5	5	4	5	4	5	5	4	4	4	5	4	5	4	5	5	5	3	5	5	5	5	4	5	4	4
61	5	3	5	5	5	5	5	5	4	5	5	5	5	5	5	4	5	5	5	5	5	5	5	5	5	5	5	5	4	4	4	5	5	4	4	4	5	4	5	5

Note: Respondents are represented across the top and Questions are represented down the left side.

Next, divide the difference by the standard deviation. This is called a *z*-score (or normal score) and shows by how many standard deviations a score falls above or below the benchmark. Lastly, convert the *z*-score to a percentile rank by using the properties of the normal curve.
   a. Analysis yielded a score of 86 percent for *z*-score to percentile rank
5. **Coefficient of Variation (CV)**: The standard deviation is the most universal way to communicate variability, but it is hard to translate. The CV makes inferring easier by dividing the standard deviation by the mean. Higher values indicate higher variability. The CV is a measure of variability, unlike the first four results which are measures of the central tendency, so it can be used in addition to the other approaches.
   a. Analysis yielded a score of 12 percent for coefficient of variation[1]

Based on the results of the general analysis above, it is apparent that the respondents were favorable toward the concepts presented in Chapters 6 and 7, indicating a high level of confidence in industry acceptance and use.

In addition to the quantitative analysis above, the respondents were given the opportunity to comment on the survey questions as a means of giving additional context to the confidence judgments made. All remarks made during survey administration expressed an appreciation of the concepts, an acknowledgment of the benefits of implementing them, and belief in a strong likelihood of future implementation and industry reception [1].

NOTE

1 For specifics on these statistical methods, see J. Lewis, and J. Sauro, *Qualifying the User Experience. Practical Statistics for User Research*, 2nd ed. (Denver, CO: Create Space Publishing, 2016).

REFERENCES

1. Corrado, Jonathan. 2017. "Technological Advances, Human Performance, and the Operation of Nuclear Facilities." PhD dissertation, Colorado State University, Fort Collins, CO, ProQuest (AAT 10258407).
2. Republished with permission of American Society of Mechanical Engineers, from Corrado, Jonathan. 2022. "Survey Use to Validate Engineering Methodology and Enhance System Safety." *ASCE-ASME Journal of Risk and Uncertainty in Engineering Systems, Part B: Mechanical Engineering* 8(2). doi:10.1115/1.4053305; permission conveyed through Copyright Clearance Center, Inc.

Index

Note: **Bold** page numbers refer to tables; *italic* page numbers refer to figures and page numbers followed by "n" denote endnotes.

Printed in the United States
by Baker & Taylor Publisher Services